ELEMENTARY ASTRONOMY

A Simple Reference Guide to Common Celestial Phenomena

for anyone interested in learning more about astronomy

Written by James N. Pierce

Good Apple
A Division of Frank Schaffer Publications, Inc.

Dedication

To my wife, Rebecca

Editors: Barbara G. Hoffman, Karen P. Hall, Christine Hood, Michael Batty
Book Design: Terry McGrath
Book Illustration: Tani Brooks Johnson
Book Production: Terry McGrath

Cover Photos courtesy of NASA, NOAO, and JPL

Good Apple
A Division of Frank Schaffer Publications, Inc.
23740 Hawthorne Boulevard
Torrance, CA 90505-5927

GA13050

Table of Contents

In the next twenty centuries, the age of Aquarius of the great year, the age for which our young people have such high hopes, humanity may begin to understand its most baffling mystery—where are we going? The earth is, in fact, traveling many thousands of miles per hour in the direction of the constellation Hercules—to some unknown destination in the cosmos. Man must understand his universe in order to understand his destiny.

— NEIL ARMSTRONG,
speech to Congress, September 16, 1969

Introduction

Open Star Cluster PHOTO COURTESY OF NOAO

This is a book about astronomy, the study of the Sun, Moon, planets, stars, galaxies, and other objects found in the sky. It is a crash course written for anyone who has never found the opportunity to learn astronomy before now—and for elementary and middle school teachers as a source of information on the basic astronomical topics that are part of a strong science curriculum.

Astronomy is the perfect introductory science: it is interesting, it is fun, and it doesn't get your hands dirty. Children are eager to learn about planets and stars, worlds much different from their own, giving their imaginations a flying start. They are amazed by the huge numbers, distances, and sizes that are encountered with every new topic, and they like being able to make predictions of moon phases, eclipses, and other events in the sky, which can then be verified by observations. In a school setting, astronomy is a natural cross-curricular topic—it develops and exercises mathematical and scientific thinking processes and stimulates the imagination. Integrating curriculum strands is easy with astronomy—it provides topics for learning activities in language arts, social studies, math, and science.

The book presents a simplified view of astronomy. Equations have been minimized, numbers have been rounded, and the diagrams can be easily reproduced on a classroom chalkboard or whiteboard, which is where most of them originated. Definitions have been placed throughout the book where useful as well as in the glossary, beginning on page 107.

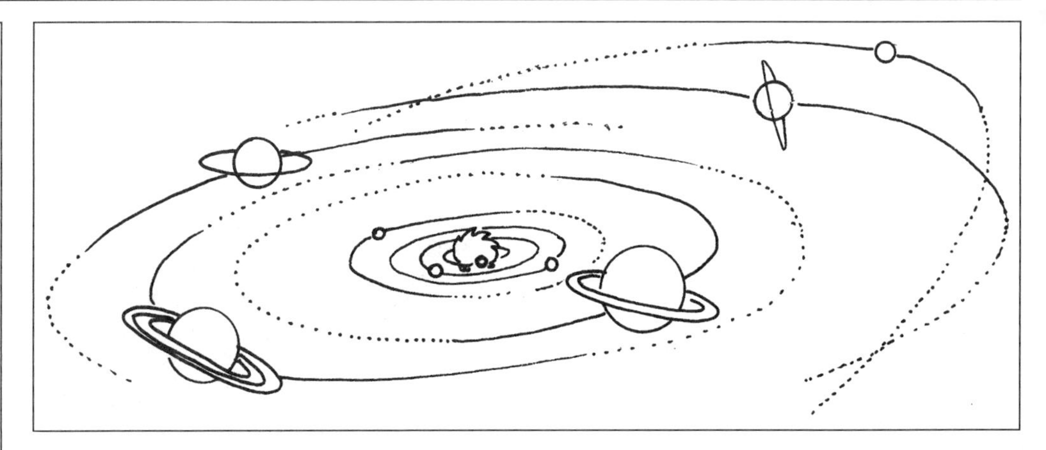

We live on the Earth, which is a **planet**. It is one of nine planets which orbit our Sun. The Sun, the planets and their **satellites**, and the **asteroids**, **comets**, and other space debris that also orbit the Sun make up the **Solar System**. The Sun is a **star**—our star—and there are many, many stars in the sky. These billions of stars, together with their planetary systems, make up the huge stellar system we call a **galaxy**. And there are billions of galaxies out there populating the universe, which includes everything there is. Although astronomy covers the entire range of topics and objects from the Earth to the **universe**, this book focuses on our Solar System to cover the astronomy topics normally taught in elementary and middle schools.

asteroid
small, rocky chunk of matter orbiting the Sun; also called a *minor planet;* most asteroids orbit in the asteroid belt

comet
mass of frozen gases revolving around the Sun, generally in a highly eccentric orbit

galaxy
an assembly of gas, dust, and typically billions of stars, all bound together by gravity

planet
body in orbit around the Sun; nine planets orbit the Sun, including Earth

satellite
natural—a moon that orbits a planet
artificial—a human-made object sent into space to orbit a planet

Solar System
the Sun and all of the objects that revolve around it

star
one of many points of light in the night sky that maintains a fairly constant position with respect to its neighbors; a massive, gaseous sphere, heated by gravitational compression until it radiates visible light; most stars generate energy by thermonuclear fusion; the Sun is a very close star

universe
the totality of known or supposed objects and phenomena throughout space

Useful Tools

Astronomy is an observational science. Experience it for yourself to better understand it. Take your time reading this book. Use it as a reference when you look at the sky.

A pair of binoculars or a telescope will help you in your observations. Most observations recommended in this book can be done easily without these tools, but if available, they will enhance your experience.

In the Appendix (page 105) you will also find information about photographing night skies. To do this, you need a 35mm camera with manual shutter controls, including a B (or bulb) setting that allows you to determine the length of time that the shutter is open.

NSES Content Standards for Space Science

Most state education standards and frameworks are written based on the National Science Education Standards (NSES). These standards are presented by grade ranges: K–4, 5–8, and 9–12. Consequently, specific grade-level study topics vary considerably from state to state and district to district. The NSES content standards published in 1996 follow. The science curriculum of your school or district will most likely be written to reflect these content standards. Check your curriculum guides to discover which space science topics you should cover at your grade level.

National Science Education Standards is available for sale from the National Academy Press, which may be reached at 2101 Constitution Avenue, NW, Box 285, Washington, DC 20055; on the Internet at *www.nas.edu*; or by calling 800-624-6242.

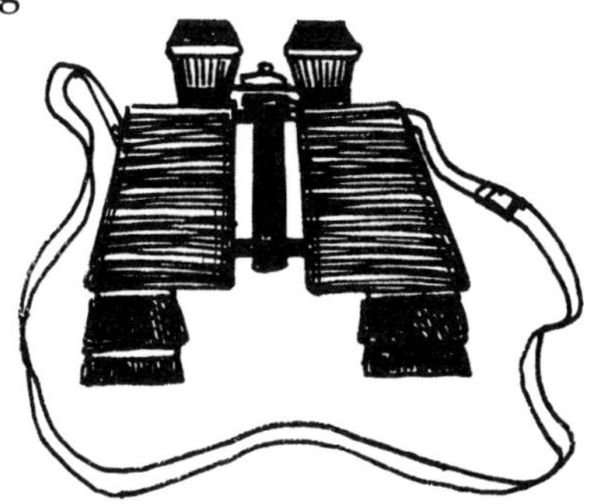

Grades K–4

All students should develop an understanding of objects in the sky and changes in Earth and sky based on regular observations of day and night skies. In K–4 emphasis is placed on developing observation and description skills, and explanations based on what students observe. They learn to identify the sequences of changes in what they observe over time, and to look for patterns in these changes. They observe the movement of an object's shadow during the course of a day, the positions of the Sun and Moon to find the patterns of movement, and record the Moon's shape on a calendar every evening over a period of weeks.

The fundamental astronomical concepts underlying the content standards are as follows:

- ✦ The properties, locations, and movements of the Sun, the Moon, and stars can be observed and described.
- ✦ The temperature of the Earth is maintained by the Sun's light and heat.
- ✦ Patterns of movement of objects in the sky can be observed. For example, although the Sun seems to move across the sky the same way every day, its path changes slowly over the seasons. The Moon's movement across the sky on a daily basis is much like that of the Sun. From day to day the observable shape of the Moon changes in a cycle that lasts about a month.

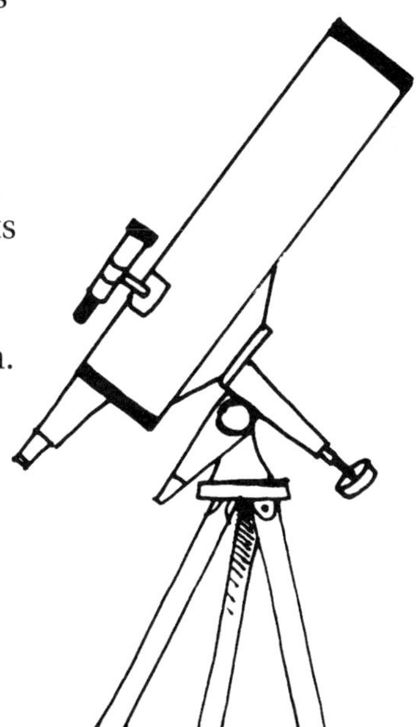

Grades 5–8

In grades 5–8, content of earth and space science focuses on developing an understanding of the Earth and the Solar System and how closely they relate as systems. The observations made in grades K–4 are the basis for constructing models to explain the relationships among the Earth, Sun, Moon, and Solar System. Students will conclude through direct observation and satellite data that the Earth is a moving, spherical planet with features that distinguish it from other planets in the Solar System. Activities that explore trajectories and orbits with the Earth-Sun and Earth-Moon systems as examples help students understand that gravity holds all parts of the Solar System together. The primary energy source for processes on the Earth's surface is energy from the Sun transferred by light and other radiation.

The fundamental astronomical concepts underlying the content standards are:

✦ The Solar System includes the Moon, the Sun, nine planets and their moons, and smaller objects such as asteroids and comets. The Earth is the third planet from the Sun. The Sun, an average star, is the central and largest body in the Solar System.

✦ Most objects in the Solar System are in regular and predictable motion. Those motions explain such phenomena as the day, the year, phases of the Moon, and eclipses.

✦ The force that keeps planets in orbit around the Sun and governs the rest of the motion in the Solar System is gravity. It alone holds us to the Earth's surface and explains the phenomenon of tides.

✦ The Sun is the major source of energy for the Earth and the phenomena on the Earth's surface, such as plant growth, winds, ocean currents, and the water cycle. The tilt of the Earth's rotational axis causes variation in the length of days and in seasonal weather in different places on the Earth's surface over the course of a year.

Grades 9–12

The study of astronomy in grades 9–12 moves from the behavior of objects in the Solar System to more abstract concepts. Students are more able to comprehend vast distances, long time scales, and nuclear reactions. By looking outward, astronomers have demonstrated that we live in a vast and ancient universe. Students are fascinated by the age of the universe and how galaxies, stars, and planets have evolved.

One of the greatest challenges of teaching astronomy in grades 9–12 is that direct experimentation relating to the concepts is difficult or impossible. Students who are not yet capable of understanding content based on abstract concepts like the space-time continuum will need concrete examples, and guidance through the multiple, logical steps needed to develop this understanding.

The fundamental astronomical concepts underlying the content standards are:

✦ One of the greatest questions in science is the origin of the universe. According to the "big bang" theory, the universe began in a hot dense state between 10 and 20 billion years ago, and the universe has been expanding ever since.

✦ Matter—primarily light atoms like hydrogen and helium—clumped together by gravitational attraction early in the history of the universe. These clumps formed countless trillions of stars. The visible mass of the universe is formed by billions of galaxies, each of which is a gravitationally bound cluster of billions of stars.

✦ Nuclear reactions—primarily the fusion of hydrogen atoms to form helium—are the means by which energy is produced in stars. The formation of all the other elements has occurred as a result of these and other processes in stars.

Chapter 1

The Earth, the Moon, and the Sun

Cave Dweller Astronomy

Let us go back in time to the days (and nights) of the cave dweller before written language. These earliest astronomers had to rely on their own observations and intelligence to figure things out. As an example, consider the cave dwellers' perceptions of the Earth, the Moon, and the Sun:

The scene opens with a group of cave men and cave women sitting in a cave (where else?) having a lively discussion about natural science. They explore their knowledge of where they live and of the brightest objects in the sky. They formulate three simple questions to compare and contrast what they know and can observe.

1. What does it look like?
2. What is its size?
3. Does it move?

To find answers to their questions, one of them goes outside to make observations and report his or her findings.

The Earth

What does the Earth look like?
The answer to this question will depend to some degree on the local terrain around the cave, but possibly the cave dweller arrives at an answer like *flat.* (Not perfectly flat, of course, but close enough for this exercise.)

What is the size of the Earth?
After some contemplation, our cave dweller decides that the Earth is rather large, and therefore, the answer to this question is *big.*

Does the Earth move?
A silly question, thinks the cave dweller, because anyone can see—or rather feel—that the Earth does not move. Just stand and look down at it—it is not going anywhere. The obvious answer is *no.*

Thus, by simple observations, the cave dwellers learn that the Earth is flat, big, and does not move.

The Moon

What does the Moon look like?
This is rather tricky because the Moon's form is not always the same—it varies from day to day.

What is the size of the Moon?
It appears rather small as it hangs in the sky—certainly much smaller than the Earth.

Does the Moon move?
Not rapidly, but yes, it does move across the sky, given enough time.

FIGURE 1.1

Scientific Findings of the Cave Dwellers

Object	Form	Size	Motion
Earth	Flat	Big	No
Moon	Varies	Small	Yes
Sun	Round	Small	Yes

The Sun

What does the Sun look like?
The Sun is generally round.

What is the size of the Sun?
The Sun appears to be similar in size to the Moon—small.

Does the Sun move?
Yes, the Sun moves gradually across the sky during the day.

Figure 1.1 summarizes the scientific findings of the cave dwellers.

If you have your students test their powers of observation, they may arrive at the same findings. Our study of astronomy will help you understand what the Earth, Moon, and Sun are really like.

Shape

In our time, we have seen photographs showing the spherical Earth floating in space. Even before humankind traveled into space, the ancient Greek astronomers deduced the Earth's shape by noting the curved shadow cast on the Moon by the Earth during a **lunar eclipse**. Only a sphere would always cast such a shadow.

Persons living on the shores of large bodies of water noticed that when ships sailed away, the bottoms of the ships disappeared from view before the tops of the masts did, implying a curved surface to the water. We know that the Earth, the Moon, and the Sun are round. As round is not a precise word, the word *sphere* (or *ball*, for young children) is a more accurate description.

We know that the Sun is a sphere because we view the Sun from different directions in space, and the Sun always presents a circular profile. The only shape that always appears circular is a sphere. Experiment holding a ball in your hands and viewing it from different angles.

The Earth is not a perfect sphere but is slightly flattened at the poles and bulging at the equator. We say that the Earth is **oblate**.

Equatorial diameter—7,928 miles

Polar diameter—7,902 miles

Difference—26 miles

The Earth's oblateness, which is the difference between the equatorial diameter and the polar diameter divided by the equatorial diameter, is about 1/300, or about one millimeter on a 12-inch globe.

FIGURE 1.2

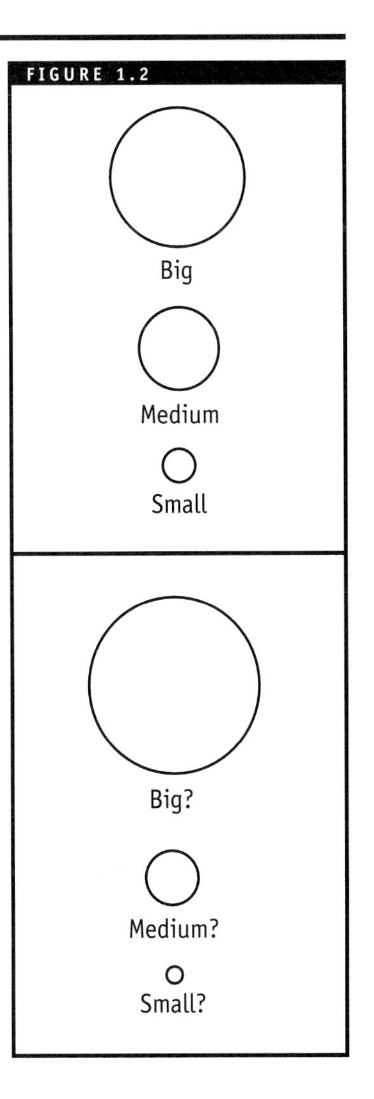

oblate
flattened at the poles

Earth Diameters

Equatorial diameter—7,928 miles

Polar diameter—7,902 miles

Difference—26 miles

Size

The cave dweller identified the sizes of the Earth, the Moon, and the Sun as *big, small,* and *small.* People living today know that the Moon is smaller than the Earth, and the Earth is smaller than the Sun, so we could rank them in size as *small, medium,* and *large.* But anyone who buys soft drinks or fries at fast food restaurants knows that the word *large* at one restaurant does not translate to the same-sized serving at another (see Figure 1.2).

PHOTO COURTESY OF NASA

Using Measurement to Compare Sizes

To make proper comparisons, use numbers. Measuring is a good basis of comparison for size. A three-dimensional object like a sphere offers several choices of dimension to measure.

You can measure the distance from the center of a sphere to the outside. That distance is the **radius (*R*)**. The plural of radius is *radii.*

You can measure the longest distance through a sphere, passing through the center. That distance is the **diameter (*D*)**.

You can measure the distance all the way around a sphere. That is the **circumference (*C*)**.

You can also measure the **surface area** (*A*) of a sphere.

Or you can measure the **volume** (*V*) of a sphere.

Typical units of measurement are miles or kilometers (squared miles, cubic kilometers, and so on). The sidebar lists the values of these five dimensions of the Earth.

As you compare different ways to measure a sphere, notice that the different properties of a sphere are related to each other. Each measuring method includes the measurement of the radius (R) as a factor.

$$D = 2R \qquad C = 2\pi R \qquad A = 4\pi R^2 \qquad V = \tfrac{4}{3}\pi R^3$$

For this reason, we need only to compare one value (numerical measurement) for each sphere, usually R or D. Comparing the radii of the Earth, the Moon, and the Sun in kilometers provides the following:

Earth ≈ 6,400 km   Moon ≈ 1,700 km   Sun ≈ 700,000 km

Comparing these numbers, you see that the Sun is the largest of these three bodies—much larger than either the Earth or the Moon. The Moon is a bit smaller than the Earth.

These numbers are large. Large numbers are best written in **scientific notation** (used for the kilometer values of volume and surface area in the sidebar).

If you measure the length of your classroom in millimeters, the measurement will be a large number. If you use bigger units, such as yards, the number will be smaller.

Miles and kilometers are normally adequate for human activities on the Earth's surface. The distances and sizes measured in astronomy are large in scale, compared to the units measured in everyday life. A unit of measurement substantially bigger than a mile is needed.

circumference
the distance around a circle or sphere

diameter
any line that joins two points of a circle and passes through its center or that joins two points of the surface of a sphere and passes through its center

Earth radius
the radius of the Earth ≈ 6,400 km

pi (π)
the ratio of the circumference of a circle to its diameter, approximately (≈) 3.141592654359

radius
the distance from the center of a circle to a point on the circle (in a sphere, the distance from the center of the sphere to a point on the surface of the sphere)

scientific notation
number expressed as a product of two factors; the first factor is a number between 1 and 10, and the second factor is a power of 10

surface area
the total area of the surface of a solid

volume
the amount of space that an object occupies

Dimensions of the Earth

Radius	3,964 miles 6,378 km
Diameter	7,928 miles 12,756 km
Circumference	24,900 miles 40,074 km
Surface Area	≈ 197,500,000 sq miles 5.06×10^8 sq km
Volume	≈ 260,900,000,000 cu miles 1.07×10^{12} cu km

1 Mile ≈ 1.6 Kilometers

≈ means approximately

Earth Radius

The unit needed is called the **Earth radius**—equal in size to the Earth's radius. If used to measure the radius of each object, the following results are obtained:

		Radius (km)	Earth Radii
Earth	≈	6,400	1
Moon	≈	1,700	0.27
Sun	≈	700,000	109

These numbers are easy to understand. The Sun has a diameter over 100 times greater than the Earth's, while the Moon's diameter is about a quarter of the Earth's.

Viewing Size Relationships

Drawing to Compare

Draw three circles to represent the true relative sizes of the Earth, Moon, and Sun—the radii of the circles would have the ratios given in the *Earth Radius* section above. If you drew a circle with a radius of one inch to represent Earth, you would draw a circle with a radius of approximately ¼ of an inch to represent the Moon, and a circle with a radius of 109 inches (about nine feet) to represent the Sun. To draw a circle that big, cut a rope the length of the radius and use it as a makeshift compass.

Making a Model to Compare

The correct relative sizes can be demonstrated by using a set of three spheres. Acquire a marble with a ½-inch diameter, a BB (0.18 inch or 0.46 cm), and a ball about five feet across.

The marble represents the Earth, the BB represents the Moon, and the five-foot ball represents the Sun. This model is a fair representation of the true relative sizes of the Earth, the Moon, and the Sun.

You need to know the distances of the Moon and Sun from Earth in order to correctly orient them in your model.

DISTANCE FROM EARTH

Moon ≈ 384,000 km	**Sun** ≈ 150,000,000 km
60 Earth radii	23,400 Earth radii

TRAVEL TIME AT 60 MPH

5.5 months	177 years

To complete the model, place the Moon (the BB) 16 inches from the Earth (the marble) and the Sun (the five-foot ball) 175 yards from the Earth.

You can construct a different model by using larger spheres. Let the Earth be a 12-inch-diameter globe. The Moon can be a baseball at a distance of 30 feet from the globe, while the Sun is the size of a baseball infield at a distance of two and a quarter miles from the globe!

If you were to drive completely around the Earth at 60 mph on a highway that allowed you to ignore the natural barriers of oceans, mountains, and other hazards, you would drive nonstop for 17 days, with no stops for sleeping, meals, or using the bathroom.

If you were to take a class on a field trip around the world, traveling only during school hours (six per day), five school days a week, your field trip would last approximately 13½ weeks.

Properties of the Earth, Moon, and the Sun

atmosphere
layer of gases surrounding an astronomical body

gas
matter composed of atoms fairly independent of each other; has a free volume and a free shape

liquid
matter composed of particles bound weakly together; has a fixed volume and a free shape

matter
the substance or substances of which any physical object consists, or is composed

solid
matter composed of tightly bound atoms; has a fixed volume and a fixed shape

temperature
a measure of the amount of energy contained in a body

TEMPERATURE SCALES

Fahrenheit scale
water boils at 212°
water freezes at 32°
absolute zero -459°

Celsius scale
water boils at 100°
water freezes at 0°
absolute zero -273°

Kelvin scale
water boils at 373°
water freezes at 273°
absolute zero 0°

Differences in the physical properties of the Earth, the Moon, and the Sun are largely due to the differences in temperature among them, the states of matter that form them, and their masses.

Temperature

Temperature is a measure of the amount of energy contained in a body. A hot body contains more energy than a cool body. We can use terms such as *hot* and *cold* to describe objects, much as we described sizes by using *big* or *small*, but the information contained in such words is limited. In order to convey more than the relative energy content of a body, it is necessary to employ some kind of temperature scale.

Temperature Scales

There are several temperature scales used today. Most North Americans are familiar with the **Fahrenheit scale**, while much of the rest of the world uses the **Celsius scale**. Both of these scales are based on the freezing and boiling points of water, which freezes at 32°F (0°C) and boils at 212°F (100°C). Large positive numbers indicate very high temperatures, while at very low temperatures, one finds negative values in each scale. Although there is no upper limit on temperature, there is a lower limit, known as *absolute zero*—approximately –459°F (–273°C). Astronomers sometimes use Fahrenheit and sometimes use Celsius, but more often use a different scale called **Kelvin**. Kelvin has its zero point at absolute zero and degrees equal in size to Celsius degrees, producing a temperature scale with no negative values, freezing at 273°, and boiling at 373°.

As you read about various astronomical objects, you will encounter different temperature scales. It is important to note which scale is being used in order to correctly interpret the temperature.

Measuring Temperature

There are many variables to consider when measuring the temperature of an astronomical body. First, you must decide where and when to measure the temperature—on the inside or outside of the object; on the surface or in the atmosphere; at the poles or at the equator; during the night or during the day. A map of the Earth documenting one day of weather demonstrates that the temperature varies with times, dates, and locations of the measurements. A single temperature will not characterize the whole planet because there is a large range of values to be reported. Similarly, other bodies in space will not necessarily have just one temperature but many.

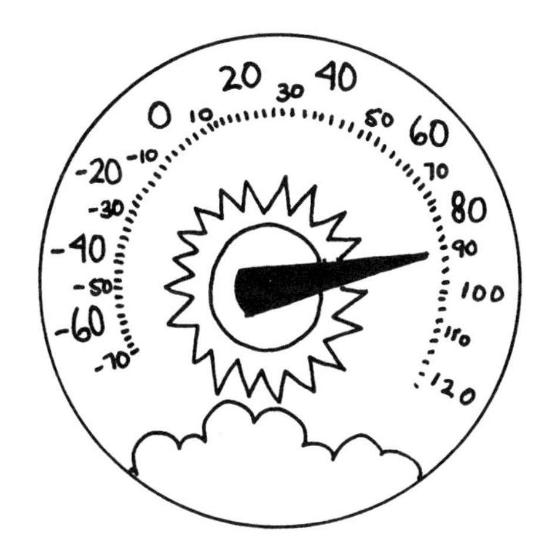

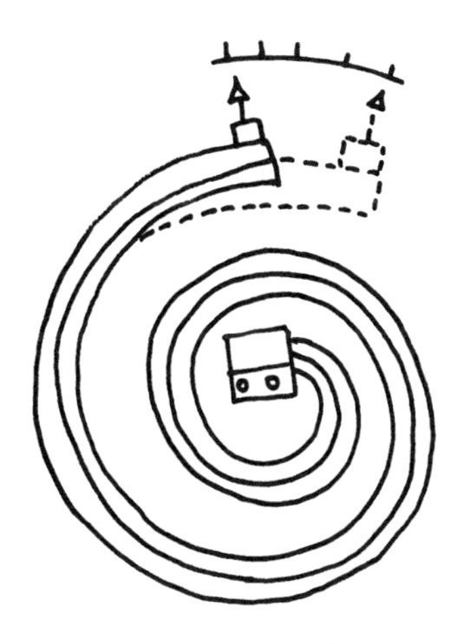

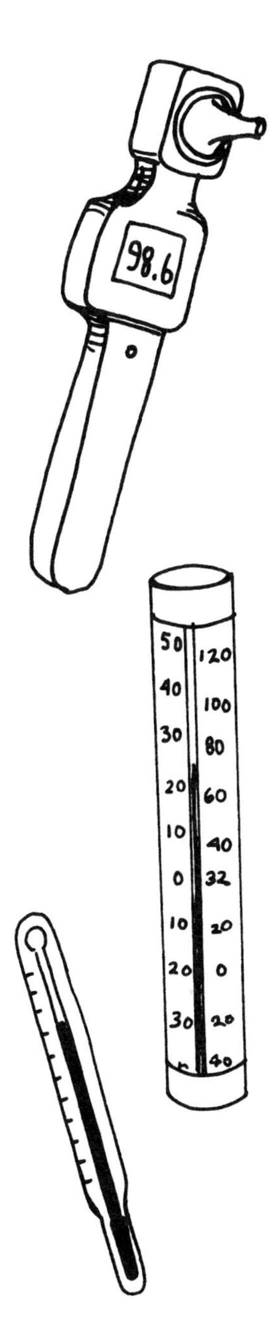

Comparing Temperatures of the Earth, the Moon, and the Sun

For the Earth, use a typical range of values for the air temperature at the Earth's surface, both in the day and at night. The daytime side of the Earth is warmer because it receives heat from the Sun. The Earth is constantly radiating energy away into space, which causes the nighttime side to cool down. The Earth's atmosphere insulates it from extreme daily temperature shifts like those found on the Moon.

The Moon warms during the lunar day and cools during the lunar night. As the Moon has no **atmosphere**, and consequently no insulating layer of air, it experiences greater extremes of temperature than the Earth. It also means that air temperature cannot be measured on the Moon. Instead, the temperature of lunar soil is measured as a basis of comparison.

The Sun is completely different from the Earth and the Moon; it is a star—a huge ball of very hot gas that radiates energy, primarily as visible light. Of course, there is no night on the Sun, and there should be no great variation over its surface as there is on the Earth. The temperature of the Sun's surface is approximately 5,800°K (9,980°F). This temperature is so high that most materials found on Earth would be vaporized. The interior of the Sun is even hotter, reaching about 15 million°K (27 million°F) at the center.

FIGURE 1.3

Temperature Comparison

	DAY	NIGHT
Earth – Surface Air	277°–310°K 39°– 99°F	260°–283°K 9°– 50°F
Moon – Soil Surface	Max. 380°K 225°F	Min. 100°K –279°F
Sun Surface	5,800°K 9,980°F	
Center	15 million°K 27 million°F	

States of Matter

Solid

A **solid** is **matter** composed of tightly bound atoms or molecules, such that their motion is quite restricted. A single piece of solid matter does not flow (although a collection of solid particles, such as sand in an hourglass, can flow), nor does it expand or contract on its own. A solid needs no container to hold it as it has a fixed volume and a fixed shape.

gas
matter composed of atoms fairly independent of each other; has a free volume and a free shape

liquid
matter composed of particles bound weakly together; has a fixed volume and a free shape

mass
a measure of the amount of matter contained in a body

matter
the substance or substances of which any physical object consists, or is composed

solid
matter composed of tightly bound atoms; has a fixed volume and a fixed shape

Liquid

A **liquid** is matter composed of atoms or molecules that are bound weakly together, allowing them to move past each other while still maintaining contact. A liquid can flow and requires a container, but it generally does not expand to fill all available space in the container. A liquid has a fixed volume and a free shape.

Gas

A **gas** is matter composed of atoms or molecules that are fairly independent of each other and move about without maintaining contact other than random collisions. Because gas particles are not bound to each other, gas can flow and also expand to fill any container. A gas has a free volume and a free shape.

Composition of the Earth, Moon, and Sun

The Earth and the Moon are both relatively cool bodies, with surfaces that are mostly solid. The solid bodies of the Earth and the Moon are molded into spherical shapes by their own gravitational forces. The two bodies are composed primarily of solid or molten rock and iron. The Earth is surrounded by a transparent blanket of air (gas), and a thin layer of liquid water covers a large portion of its solid surface. The Earth's gravity creates a lid to keep our atmosphere in place on our planet.

The Sun is a gas, also molded into a spherical shape by its own gravitational forces. The Sun's gravity supplies the lid that keeps its gases in place around it.

If gravity could be turned off, the gases of the atmosphere of the Earth and the entire Sun would drift away into space, while the solid parts of the Earth and the Moon would remain.

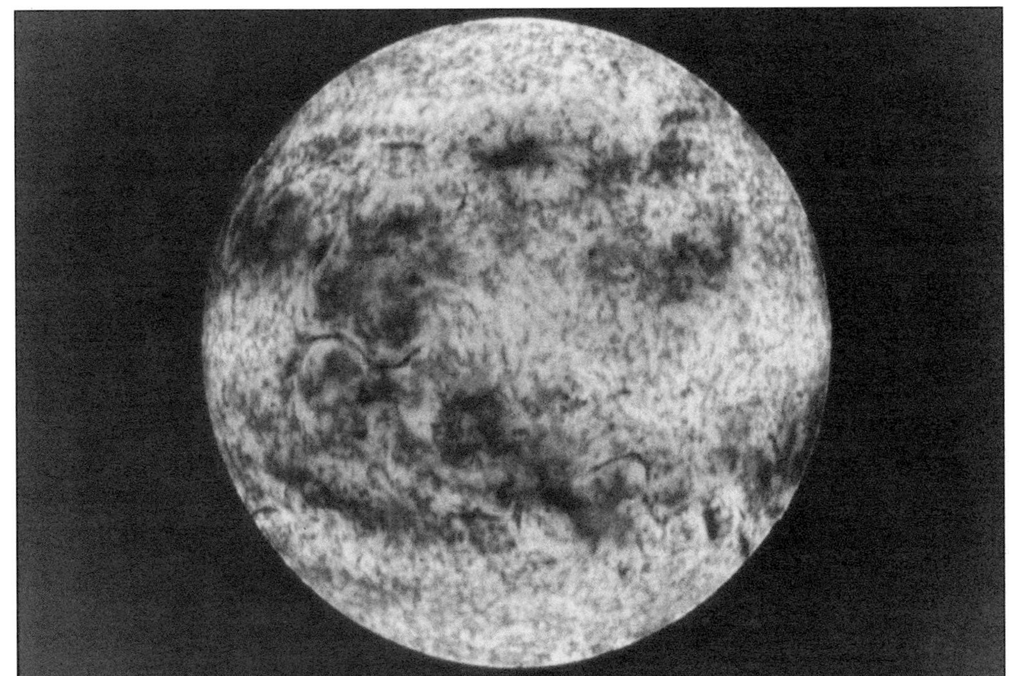

The Sun

PHOTO COURTESY OF NOAO

Mass

The physical properties of each of these bodies are a result of their respective masses as well as temperature and form.

Mass is a measure of the amount of matter contained in a body.

The Earth has a large mass and enough gravity to hold an atmosphere, making it an ideal planet for life.

The Moon's mass is so low that its gravity is too weak to hold any atmosphere at all. Human visitors to the Moon must bring their own supplies of oxygen and wear space suits to give their bodies a normal environment.

The Sun's mass is so great that its gravity has compressed and heated its matter to the point where solids and liquids cannot exist. The high temperatures that result cause the Sun to radiate light. The Earth intercepts some of this light, the ultimate source of most of the energy used by humans (and plants and animals) on this planet.

The fundamental difference between a planet and a star is mass—bodies of sufficient mass become stars, while those with lower mass will be planets. The Earth, Moon, and Sun are probably typical objects, with many similar bodies found throughout the universe.

Chapter 2
Gravity and Orbits

Gemini Astronaut PHOTO COURTESY OF NASA

Bodies in space move in paths called orbits. These orbital paths are determined by gravity. In describing the motion of bodies in space (and bodies on Earth), we use several specific terms.

Position

The **position** of an object describes its location with respect to a reference point. Some examples of position: *Your friend sits **two feet to your left;*** and *Your home is **three miles east and four miles south of the intersection of two highways**.* In both examples, the position indicates both the direction and the distance of an object from a reference point.

Velocity

Objects can move, which means that they can change position. An object will move in some direction at some rate of speed. **Speed** is the distance covered per unit of time by an object in motion. Speed in a specific direction is called **velocity**. For example, a car moving at 35 miles per hour (mph) has a velocity of *35 mph to the north* and a speed of *35 mph.*

Vector

A **vector** is a physical quantity that is described by both magnitude (size) and direction. Position and velocity are both vectors; speed is not.

Acceleration

The rate of change of velocity of an object is called **acceleration**. Changes in the speed or direction of motion are changes in the velocity of the object. Increasing speed is called acceleration, and decreasing speed is called *deceleration*, or *negative acceleration*. For example, you change the velocity of your automobile by stepping on the gas pedal, braking, or turning. Each of these produces acceleration.

Velocity and acceleration may have values of zero. An object with zero velocity is not moving. An object with zero acceleration is not changing its velocity. Velocity and acceleration are separate, independent quantities. Either one, both, or neither may be zero at a given time.

Force

Force has a simple meaning in science—most textbooks describe force as a push or pull. Make an object move and ask yourself how you did it. The answer is that you had to apply some kind of force to the object. If you move a pencil by pushing it across a table, that motion is the result of force (the push). If you drop an eraser, the force of **gravity** is involved. If you use a magnet to pick up a paper clip, magnetic force is at work. In dry weather, your bad hair days may be caused by electrostatic forces.

Forces of any kind can cause objects to move. But before modern science, people believed forces were *necessary* for motion to exist: if left to themselves, objects would always come to rest. You can demonstrate this by pushing a book across a table; it will always stop (unless it runs out of table first). The book will not keep moving unless you continue to push it.

Objects on the Earth were believed to move in straight lines. The trajectory of a cannonball was thought to resemble two sides of a right triangle, as in Figure 2.1. On the other hand, objects in the sky could have curved paths, being celestial rather than terrestrial objects.

It was not until Isaac Newton produced his famous laws of motion that simple movement of objects was satisfactorily explained. These three laws incorporate our ideas about forces and their connections with motion. They apply to terrestrial as well as celestial objects.

FIGURE 2.1

Perceived trajectory of a cannonball prior to 1687 AD

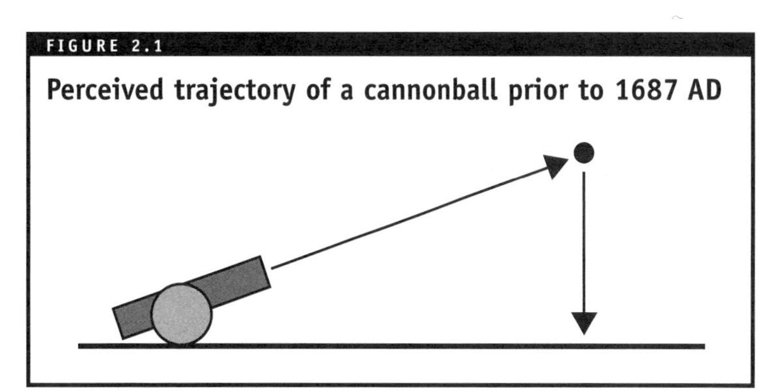

acceleration
the rate of change of velocity of an object

force
a push or pull

gravity
the attractive force that acts between all objects in the universe; on Earth, it is the force that pulls us down and holds us onto the planet

position
location with respect to a reference point

speed
the distance covered per unit of time by an object in motion

vector
a physical quantity that is described by both magnitude and direction

velocity
speed in a specific direction

Newton's Laws of Motion

acceleration
the rate of change of velocity of an object

force
a push or pull

friction
surface resistance to relative motion, as of a body sliding or rolling

gravity
the attractive force that acts between all objects in the universe; on Earth, it is the force that pulls us down and holds us onto the planet

kilogram
a metric unit of mass approximately equal to the mass of 1000 cubic centimeters of water; on the surface of the Earth a mass of one kilogram has a weight of about 9.8 newtons and about 2.2 pounds

mass
a measure of the amount of matter contained in a body

newton
a metric unit for measuring weight and other forces; the amount of force required to give a mass of one kilogram an acceleration of one meter per second per second ($1\ m/s^2$)

pound
a British unit for measuring weight and other forces; the amount of force required to give a mass of one slug an acceleration of one foot per second per second ($1\ ft/s^2$)

slug
a British unit of mass equivalent to approximately 14.6 kilograms; on the surface of the Earth a mass of one slug has a weight of about 32.2 pounds

weight
a force; the magnitude of which depends upon the mass of the object and the gravity exerted upon it

The First Law

A body at rest remains at rest, and a body in motion travels in a straight line at constant speed unless acted upon by a net ***force.***

Example: A book on a table does not move unless it is pushed. A book sliding across a frictionless table would move in a straight line without slowing down. On a real table, the force of **friction** slows the book to a stop.

The Second Law

A net force acting on a body produces an ***acceleration*** *that is proportional to the force* ($a \approx F$) *and inversely proportional to the body's mass* ($a \approx 1 \div m$)*. Together,* $a = F \div m$ *or* $F = ma$.

The **mass** of a body is a measure of the amount of matter in the body. More massive objects respond more slowly to a given force and require a greater force to achieve the same acceleration.

Example: A fully loaded truck needs more time (and gas) to get to cruising speed than an empty one.

The Third Law

For every action there is an equal and opposite reaction.

Example: If you kick a rock, the rock "kicks you back"—it exerts a significant force on your toe. A rocket moves by squirting exhaust gases backward through a nozzle; these gases exert an opposing force on the rocket, pushing it forward.

The Force of Gravity

On the Earth, the force of **gravity** accelerates objects downward, toward the center of the Earth. On the surface of the Earth, the acceleration of gravity—called g— has a fixed value of $32\ ft/s^2$. This means that a falling object increases in speed by 32 feet per second for every second it falls. (The metric equivalent is $9.8\ m/s^2$.)

Under Newton's second law ($F = ma$), an object of mass m will feel a downward gravitational force of magnitude $F = mg$.

The Force of Weight

Weight is force. The weight of an object depends on its mass and the gravity exerted upon it ($w = mg$). Gravity is discussed above. Mass is a measure of the amount of matter contained in a body (see page 17). The mass of a body remains unaffected by changes in its location. For example, the mass of a bowling ball would be the same on the Earth, on the Moon, or in space.

Weight is a force—an interaction between two objects. The weight of a bowling ball on the Earth's surface is the force that opposes the Earth's gravitational pull on the ball. The weight of the same bowling ball on the Moon's surface is different, because its weight is the force that opposes the Moon's gravitational pull on the ball. We know that the Moon's gravitational pull is low (see page 73).

In space, the bowling ball would be attracted by the Earth, the Moon, and other bodies, but with no force opposing these gravitational pulls, the ball would be weightless.

Measuring Weight

When you step on a bathroom scale, the Earth's gravitational force pulls you downward onto it. Inside the scale a spring is compressed by this force and turns the dial on the scale. Thus the bathroom scale measures the force of Earth's attraction for you—it measures your weight.

Since your weight and mass are related ($W = mg$), you can figure out your mass by dividing your weight by g. Mass is described in units called **slugs**.

If your weight is 128 pounds, then your mass is $128 \div 32 = 4$ slugs. In the metric system, your mass would be 58 **kilograms**, and your weight would be $58 \times 9.8 = 569$ **newtons**.

Mass and weight are related by the acceleration of gravity, **g**.

The Value of g

The value of g can be variable (when italicized). There are factors that affect its value, such as the mass (***M***) of the body doing the attracting and the distance (r) from the center of mass of that body. If you are standing on the surface of the Earth, then the distance is the radius of the Earth (**R**).

The equation for g contains one other factor, called **G**—the gravitational constant—a number whose value depends on the system of units being used in the equation.

The equation $g = GM \div r^2$ shows that the acceleration of gravity is higher for large mass but lower for large radius. The acceleration of gravity on the surface of the Earth is one g. On the surfaces of the Moon and the Sun it would be ⅙g and 28g, respectively. A person standing on the Moon would weigh only ⅙ of what he or she normally does on Earth. Anyone bold enough to stand on the Sun would hardly be able to move, with a weight 28 times Earth weight.

Gravity acts everywhere. **Newton's law of gravity** says that for every two masses there is an attractive gravitational force, with each mass pulling on the other. Figure 2.2 shows two such masses, M and m, separated by a distance r, and pulling on each other with a force F. The magnitude of the force is given by the equation shown.

Equations

gravitational force
$F = mg$

force = F

g acceleration of gravity on Earth
$g = 32$ ft/s²
$g = 9.8$ m/s²
$g = GM \div R^2$

G the gravitational constant—a number whose value depends on the system of units being used in the equation

m = mass
$m = W \div g$

Newton's second law
the relationship between acceleration, force, and mass
$a = F \div m$ or $F = ma$

R the radius of the Earth
3,964 miles
6,378 kilometers

weight
$W = mg$
on the surface of a planet, weight equals mass times acceleration of gravity

Relative Masses

	Earth	Moon	Sun
Mass	1	0.0123	333,000
Radius	1	0.272	109
Gravity	1	0.166	28

Newton's law of gravity
For every two masses there is an attractive gravitational force, with each mass pulling on the other.

FIGURE 2.2

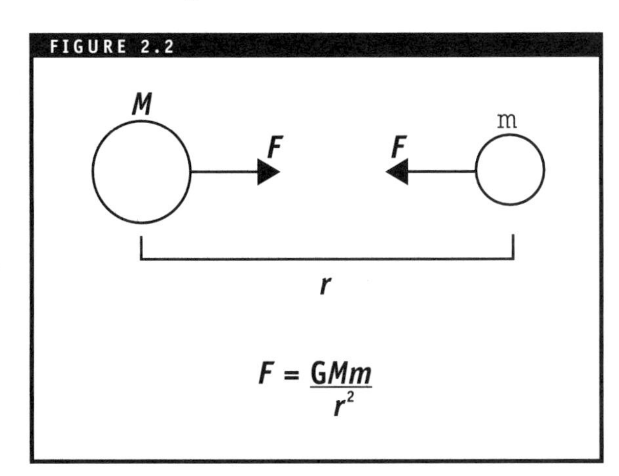

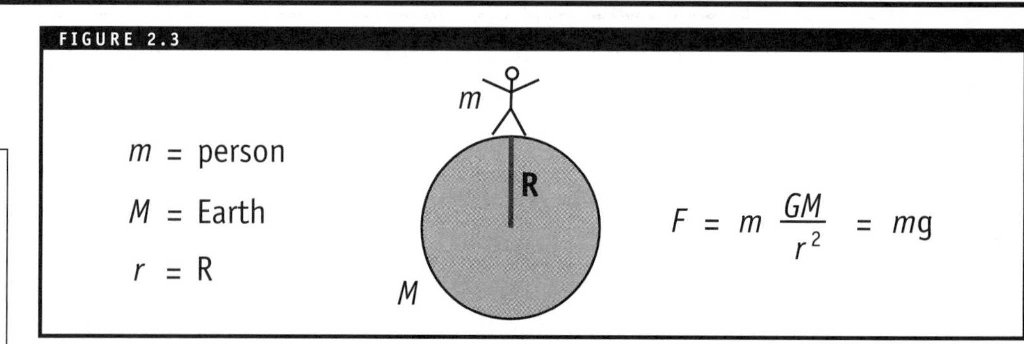

The equation $F = GMm \div r^2$ also applies if the two objects are touching. In Figure 2.3 we see that the force $(F = mg)$ is the weight of the person. This equation could use the masses of any two things, such as two people, the Earth and the Moon, or the Earth and the Sun. In each case, the equation predicts the gravitational force existing between the two masses.

Newton's laws of motion

The first law
A body at rest remains at rest, and a body in motion travels in a straight line at constant speed unless acted upon by a net force.

The second law
A net force acting on a body produces an acceleration (***a***) that is proportional to the force (***F***) and inversely proportional to the body's mass (***m***).
$a = F \div m$ or $F = ma$

The third law
For every action there is an equal and opposite reaction.

Newton's law of gravity
For every two masses there is an attractive gravitational force, with each mass pulling on the other.

gravitational force
$F = mg$

Orbits

According to Newton's law of gravity, the Sun pulls on the Earth and the Earth pulls on the Sun. Why then doesn't the Earth fall into the Sun? After all, if you let go of a tennis ball, it falls to the ground because the Earth pulls it downward.

Investigate the motion of a tennis ball by throwing it horizontally. The thrown ball moves away from you and curves downward to land on the ground. The ball does not fall straight down when it leaves your hand because it already has a horizontal **velocity**. The downward curve that you observe is the result of the ball's initial horizontal velocity and the downward acceleration produced by the Earth's gravity.

If you throw the ball harder, it travels farther before it hits the ground, but it is still accelerated downward by gravity. Suppose you could throw the ball so hard that its path curved downward at the same rate that the Earth curved away beneath it, due to its spherical shape (see Figure 2.4). The ball would travel all the way around the Earth without striking the ground. The path of the ball would then be called an orbit.

In the same manner, the Earth orbits around the Sun. As the Sun pulls on the Earth, it changes the Earth's velocity. The Earth *falls* around the Sun, never hitting it, because its speed is sufficient to keep it in orbit (see Figure 2.5).

acceleration
the rate of change of velocity of an object

centrifugal force
a reaction force to centripetal force

centripetal force
the force acting upon a body moving along a curved path that is directed toward the center of curvature of the path and constrains the body to the path

velocity
speed in a specific direction

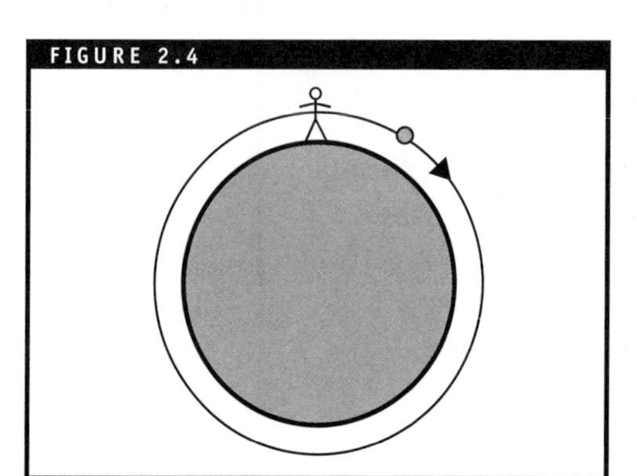

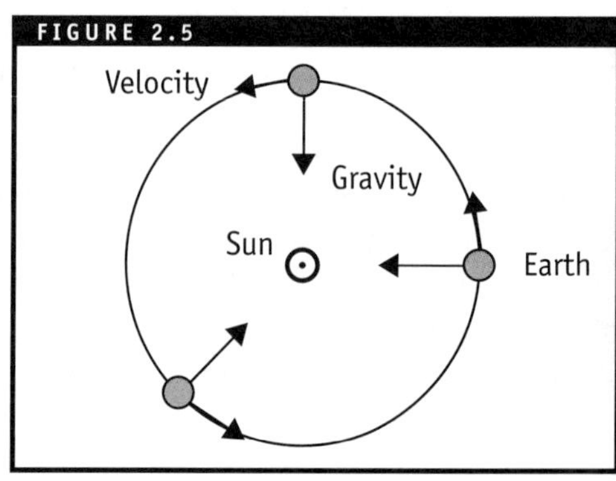

Centripetal Force

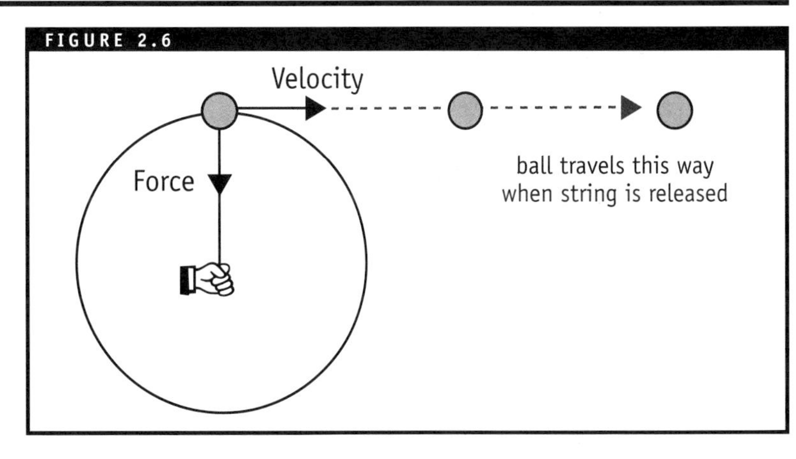

You can demonstrate the effect by attaching a string to your tennis ball and twirling the ball around your head while holding onto the string. The string supplies a **centripetal force,** which pulls the ball toward the center of the orbit where your hand holds the string. This centripetal **acceleration**, along with the initial velocity you gave the ball creates the ball's orbital motion. The ball never travels inward along the string even though that is the direction of the force. Its velocity is constantly modified by the acceleration produced by the force, and it travels in a circle around you. If you release the string, the force stops acting on the ball, the acceleration of the ball ceases, and the ball flies off in a straight line, as shown in Figure 2.6.

The Sun's gravity is a centripetal force, and it produces a centripetal acceleration of the Earth. This acceleration, together with the initial velocity of the Earth, creates the Earth's orbital motion.

Centrifugal Force

As you twirl the ball on the string, you will notice that the ball seems to exert an outward force along the string, as if it were trying to get away. This force, called a **centrifugal force**, is a reaction force (see Newton's third law of motion) to the centripetal force on the ball. There is no real force pulling the ball outward along the string, and in this sense, the centrifugal force is an illusion.

If there were such a centrifugal force, the ball would fly radially outward (or straight out) along the direction of the string when you let go, rather than moving as shown in Figure 2.6. The tangential movement illustrated is the **velocity** that the ball will have when the centripetal force acting on it is removed and the resulting centripetal acceleration disappears.

Another example of centrifugal force occurs when you turn a sharp corner in your car. When you turn left, the passengers seem to lean to the right, as if a force were pushing them toward the outside of the curve. But there is no such force. In reality, the car is turning (accelerating) due to the force of the road against the tires, and the passengers are attempting to continue moving straight ahead. When the car pushes on the passengers to make them follow the curve of the road, it appears to them as if they are being pushed against the car by a (non-existent) centrifugal force.

Try the following experiment: Get a plastic bucket or pail with a handle and put about an inch of water in it. Then stand holding the pail down at your side and move your arm and the pail rapidly in a vertical circle, such that the pail turns upside down. If you move your arm fast enough, the water does not spill out. To demonstrate how the Earth could fall into the Sun, stop the bucket at the top of the arc. You may prefer to substitute a tennis ball or several table tennis balls for the water.

FIGURE 2.7

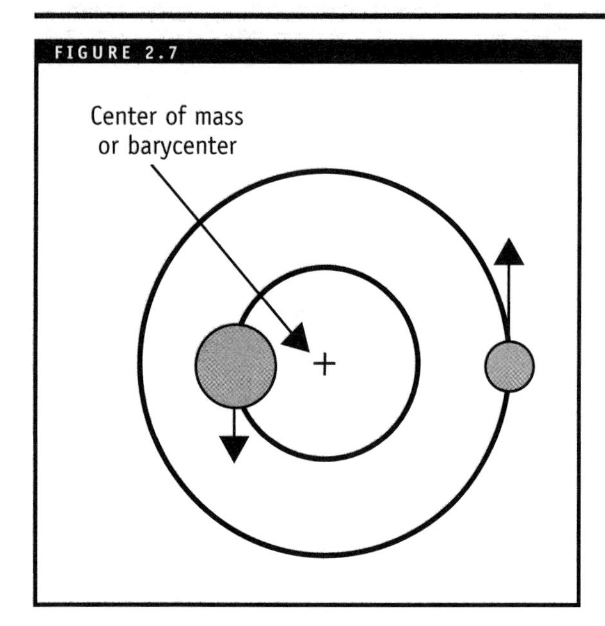

Barycenter

In a system of two bodies, neither will be absolutely stationary at the center of the orbit. Instead, both objects orbit around their common center of mass, a point called the **barycenter**, shown in Figure 2.7. The barycenter lies closer to the more massive object. In the Earth-Sun system, the barycenter is inside the Sun. In the Earth-Moon system, the barycenter is inside the Earth.

Orbital Shapes

The shapes of orbits vary. In the case of the tennis ball on the string, the shape of the orbit was a circle, because the string had a fixed length. In situations in which gravity provides the centripetal force, there is no fixed center point, and the distance between the orbiting body and what it is orbiting may vary. The shape for most orbits is an ellipse.

An **ellipse** is a geometric curve that can be described as a flattened circle. To draw an ellipse, mark two points called foci (one **focus**, two **foci**). Stick a thumbtack in each focus, connect a length of string between the two tacks, and trace out the ellipse with a pencil, as shown in Figure 2.8.

The shape of the ellipse you draw will depend on the length of the string (***L***) and the separation of the two foci (***F***). The length of the ellipse (***A***) will be approximately equal to the length of the string. The degree of flattening is measured by the **eccentricity** (***e***), which is the ratio of F to L ($e = F \div L$). The eccentricity varies from 0 to 1 as the flattening increases.

Examples of three ellipses with the same length but different eccentricities are shown in Figure 2.9. The first one has an eccentricity of 0 and is a circle. Note how the separation of the foci increases with the eccentricity.

FIGURE 2.8

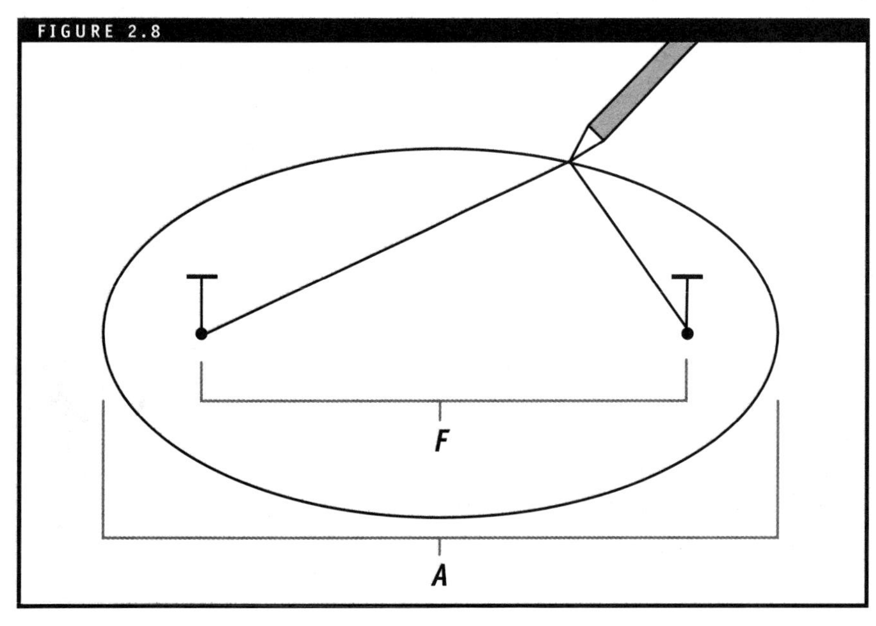

In a two-body system, both bodies would describe ellipses around the barycenter, with the barycenter located at one focus of each ellipse.

You can get a simpler picture of the motion by holding one body stationary at one focus, while the other body travels around the ellipse. In the case of the Earth-Sun system, the Sun lies at one focus, and the Earth travels around the ellipse. In doing so, the Earth changes its distance from the Sun.

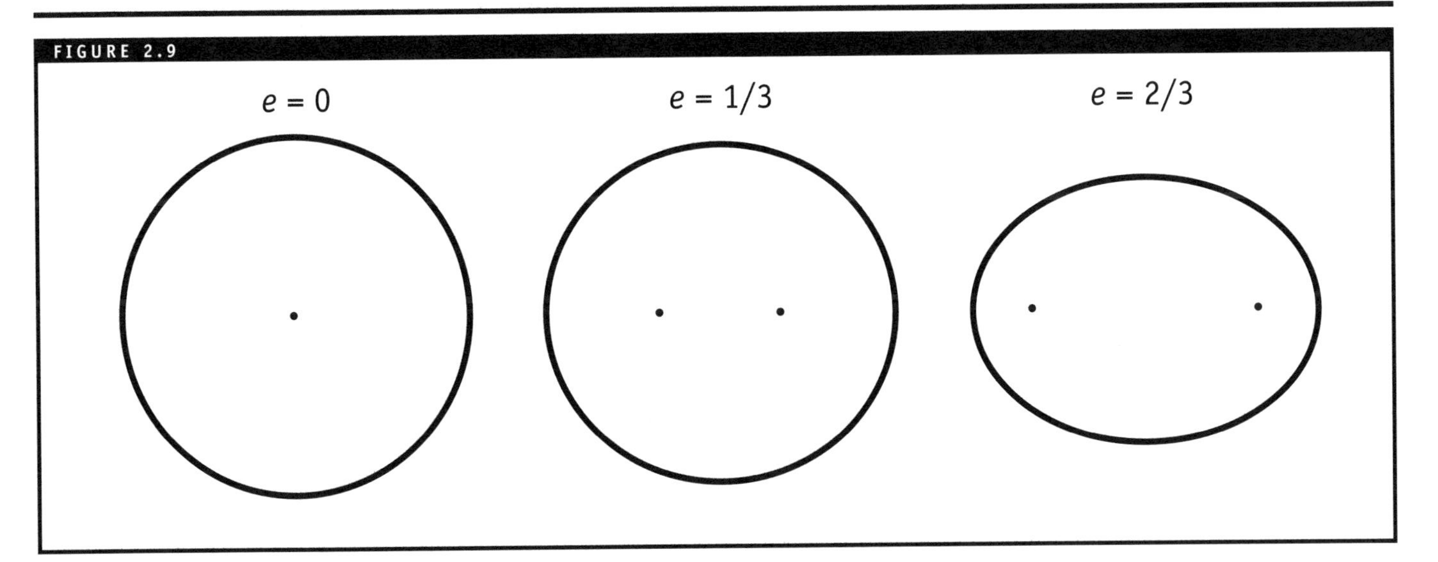

Its closest approach is called **perihelion,** while its farthest distance is **aphelion**.

These distances can be easily calculated for an orbit from the eccentricity (e) and the average distance (*a*) as follows:

Perihelion: $r_{min} = a(1-e)$

Aphelion: $r_{max} = a(1+e)$

We define the average distance of the Earth from the Sun (150 million kilometers) as one **astronomical unit (AU)**.

When we combine the value of one AU with Earth's orbital eccentricity of 0.017, we find that the Earth's distance from the Sun varies between 0.983 AU at perihelion and 1.017 AU at aphelion—not a huge variation.

Of the three ellipses shown in Figure 2.9, the Earth's orbit most closely resembles the first one (the circle).

You might think that this small variation in the Earth-Sun distance causes our seasons on Earth—summer at perihelion and winter at aphelion—but this is not the case. In fact, the Earth is closest to the Sun in January and farthest in July! The actual cause of seasons will be discussed in the next chapter.

Another quantity that varies around the orbit is the speed of the orbiting body. In general, the orbital speed will be higher when the two bodies are closer together (near perihelion) and lower when they are farther apart (near aphelion).

aphelion
the point in an object's orbit around the Sun where it is farthest from the Sun

astronomical unit (AU)
equal to about 93 million miles (150 million kilometers), the mean distance of the Earth from the Sun

barycenter
the center of mass in a system of two orbiting bodies

eccentricity
the degree of flattening of an ellipse

ellipse
a geometric curve in the shape of a flattened circle; the shape of planetary orbits

foci
plural of *focus* (one focus, two foci)

focus
one of two fixed points used to create an ellipse

perihelion
the point in an object's orbit around the Sun where it is closest to the Sun

Chapter 3
Earth's Motions and Seasons

Geographic South Pole PHOTO BY LYNN TEO SIMARSKI, NSF

Revolution

The Earth travels in a nearly circular orbit around the Sun. We call this motion **revolution** and say that the Earth *revolves* around the Sun.

To describe the direction of the revolution of an orbiting body, use one of two words—*clockwise* (CW) or *counterclockwise* (CCW). These direction words should be familiar to those who have a clock with moving hands (see Figure 3.1).

If you look at Figure 3.1 from behind the page (through the paper), the two directions are the reverse of what you see from the front of the page. To insure that you always view orbits from the same direction, and to avoid specifying your viewpoint each time you use the direction words *clockwise* or *counterclockwise,* adopt the convention of describing the orbit from the northern side—that is, the side from which you can see most of the Earth's northern hemisphere. With this convention, you can say that the Earth's revolution is *counterclockwise.*

FIGURE 3.1

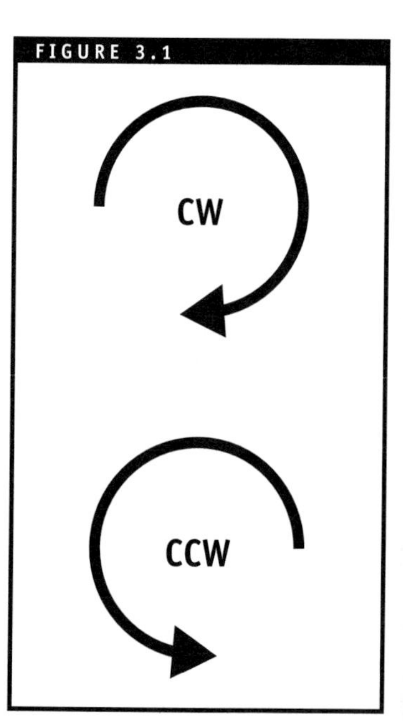

Rotation

In addition to its revolution, the Earth has another motion—it spins on its **axis**, in the manner of a spinning top, only much more slowly. This motion is called **rotation**—the Earth *rotates* on its axis.

Note that although rotation and revolution refer to two distinct motions of the Earth, they are often confused with each other because the words look and sound somewhat alike.

FIGURE 3.2

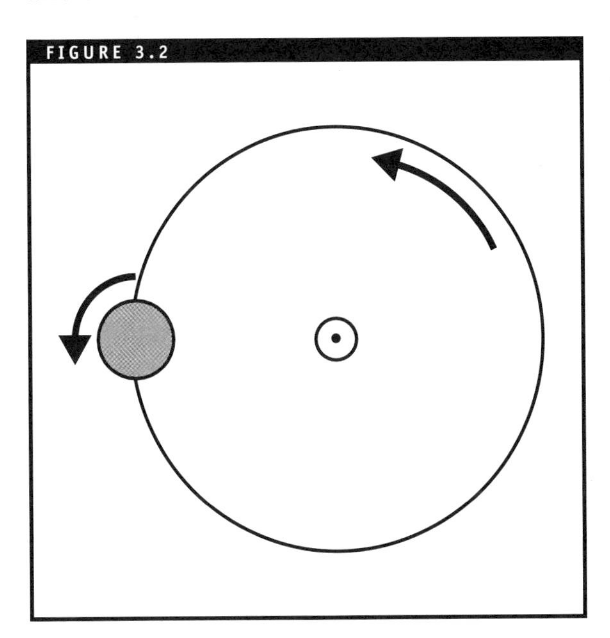

By using the above convention, we can say that the direction of the Earth's rotation is counterclockwise, the same as its revolution, as shown in Figure 3.2.

You may wonder why the Earth rotates and revolves in the same direction. The motions of the Earth (and the other planets) were established during the time of the formation of our Solar System (see Chapter 6). They have not altered substantially since then.

The Earth's rotation on its axis defines several special points on the Earth's surface, as shown in Figure 3.3. The two points where the axis intersects the surface of the Earth are called the **North Pole** and the **South Pole**. If you locate all points on the Earth's surface that are equidistant from both poles, you will identify the **equator**.

FIGURE 3.3

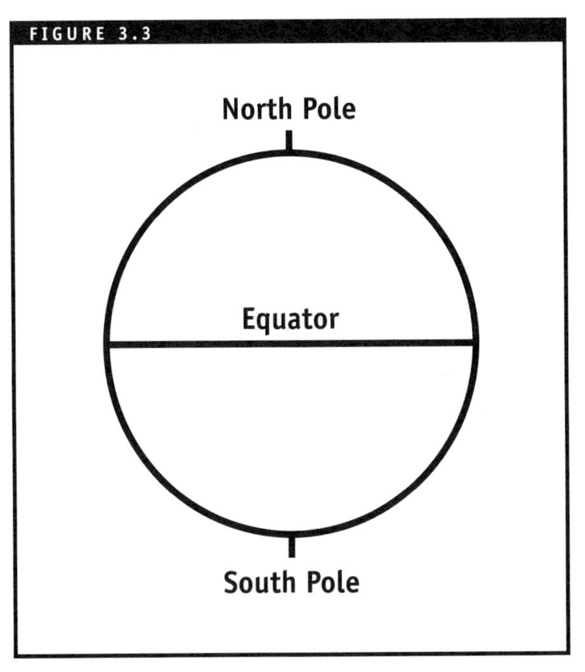

Note that the equator and the poles are defined by the rotation of the Earth. If the Earth did not rotate, you would not be able to identify these special points—at least, not in the same manner. Therefore, as you explain these topics, discuss rotation first, followed by the poles and equator.

axis
the line on which a rotating body turns

equator
all points on the Earth's surface that are equidistant from both poles

North Pole
the northern point where the Earth's rotational axis intersects its surface

revolution
the motion of a body that is orbiting (revolving) around another body

rotation
the motion of a body that is spinning (rotating) on an axis through the body

South Pole
the southern point where the Earth's rotational axis intersects its surface

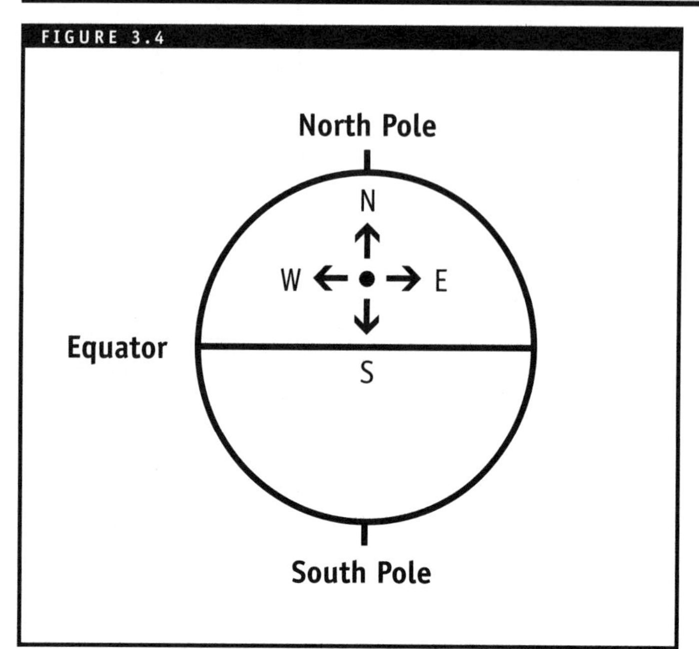

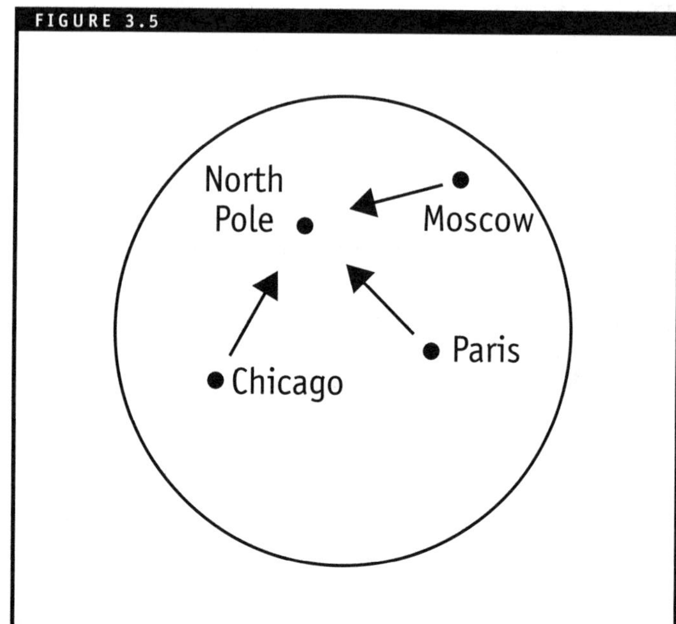

meridian
an imaginary line that runs from the north celestial pole (NCP) through the zenith to the south celestial pole (SCP), dividing the observer's sky into eastern and western halves

obliquity (tilt)
the angle between a planet's rotational axis and its orbital axis

sidereal day
the interval between successive alignments of the Earth's axis, the observer, and a fixed star—about $23^h56^m4^s$

sidereal time
the system of time based on the Earth's rotation in relation to fixed stars; because one sidereal day is $23^h56^m4^s$ long, sidereal time flows more rapidly than solar time

solar day
the interval between successive alignments of the Earth's axis, the observer, and the Sun—24 hours

Directions

Now that the poles and the equator have been introduced, we can talk about directions on the Earth. You are familiar with the terms *north, south, east,* and *west.* North is the direction of the North Pole. South is the direction of the South Pole, and east and west are parallel to the equator, as shown in Figure 3.4. All four directions lie in the plane that is tangent to the Earth's surface at the point in question. Even though we would actually have to follow the curve of the Earth to get to the North Pole, we define *north* as the direction in which we would start walking to get there.

Take a globe and choose a city. It is easy to establish which direction is north from that city. Now choose several other cities and find north for each. Note that people pointing north from each city would not all be pointing in the same direction (see Figure 3.5)—a consequence of living on a spherical Earth.

Day and Night

The Sun shines on the Earth, illuminating half of its surface. The time spent in the illuminated half is called day; time spent in the darkened half is called night. Rotation of the Earth causes day and night to alternate (for most locations) as we are carried into and out of the darkness by the turning Earth. Rotation causes the Sun to appear to move across the sky, much as your surroundings seem to move if you twirl around. This apparent motion of the Sun is the basis for keeping time. Conversely, knowing the time tells where the Sun is in the sky. Abbreviations for morning and afternoon, *a.m.* and *p.m.*, stand for *ante meridiem* and *post meridiem.* These tell whether the Sun is before (*ante*) or after (*post*) the **meridian**, the imaginary line that passes overhead in a north-south direction and divides the sky into an eastern half and a western half.

FIGURE 3.6

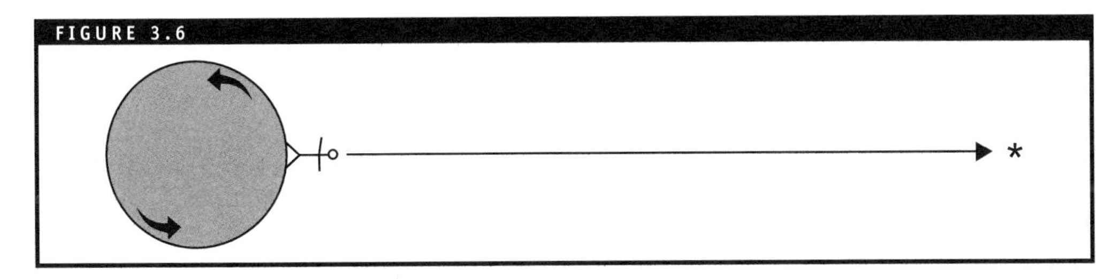

Sidereal Day

How long is one day? This interval must be carefully defined, for it can have different meanings. The word *day* could refer to a complete rotation (daytime and nighttime) or only to the daytime. Picture the Earth as shown in Figure 3.6, with a particular star directly above the observer. As the Earth rotates, the star will move slowly across the observer's sky. The length of time that passes before the star is again overhead for the observer is an interval of time we called one **sidereal day**, which is equal to 23 hours, 56 minutes, and 4 seconds ($23^h56^m04^s$).

Solar Day

Suppose the Sun were in place of the star in Figure 3.6. As the Earth turns, the Sun appears to move across the sky. The length of time it takes before the Sun is again overhead for the observer is an interval of time called one **solar day**, and is equal to 24 hours. Since humans prefer to coordinate activities with the position of the Sun in the sky, the solar day is used as the basis of our timekeeping system, and our clocks measure solar time. Thus, the Earth's rotation period is one day.

Sidereal Time

If you are interested in observing certain stars, you might want to use **sidereal time** (based on the sidereal day) to mark the Earth's rotation. Sidereal time is used primarily by astronomers.

FIGURE 3.7

Obliquity

The Earth rotates around an imaginary axis through the North and South Poles. You can also consider the Earth as revolving around another imaginary axis through the center of the orbit and perpendicular to the plane of the orbit. Thus, we have two axes (pronounced ak´sees)—a rotational axis and an orbital axis.

The two axes are inclined to each other by about 23½ degrees, an angle referred to as the **obliquity**, or sometimes as the tilt. Figure 3.7 shows a typical globe, with its rotational axis tilted by 23½ degrees to the perpendicular, which in turn represents the orbital axis. The tabletop is parallel to the plane of the Earth's orbit.

Equinoxes and Solstices

The time required for each revolution of the Earth around the Sun is one year—called the **revolution period**. In Figure 3.8, the Earth is in its orbit around the Sun, with the rotational axis oriented vertically on the page and the orbital axis angled by 23½ degrees with respect to it.

FIGURE 3.8

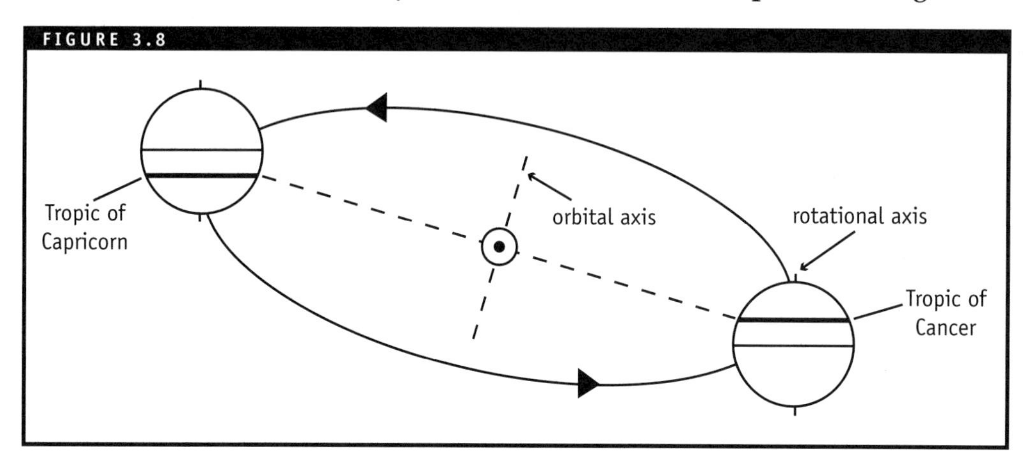

As the Earth orbits the Sun, the rotational axis does not change its orientation. It remains pointed in the same direction in space.

The orientation of the Sun with respect to the Earth's equator does seem to change. Over the whole orbit the Sun appears to move alternately northward and southward in the sky. For the Earth on the left, the Sun appears to be south of the equator and would be seen directly overhead at a point in the southern hemisphere. For the Earth on the right, the Sun appears to be north of the equator and would be seen directly overhead at a point in the northern hemisphere. Due to the Earth's tilt, the Sun appears to move as far south as 23½ degrees south latitude, a circle on the Earth called the **tropic of Capricorn**; and it appears to move as far north as the **tropic of Cancer**, at 23½ degrees north latitude.

FIGURE 3.9

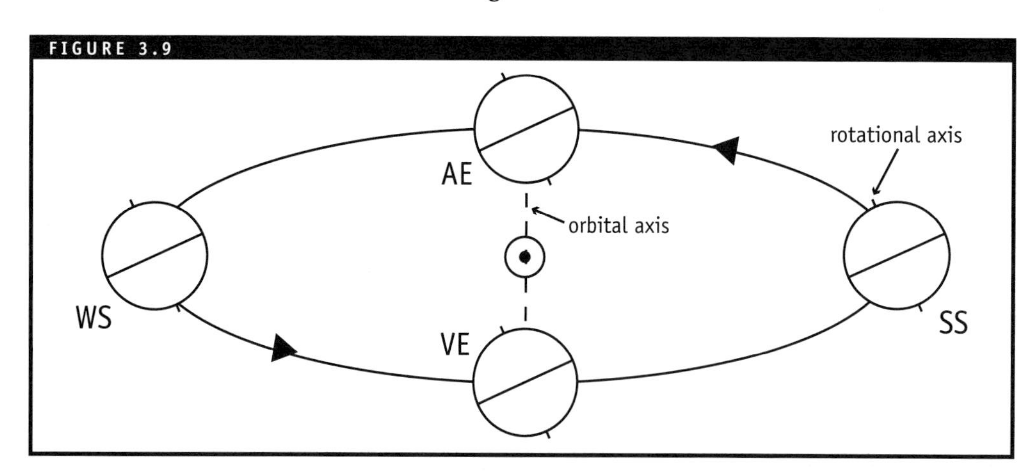

Figure 3.9 is a view of the Earth-Sun system in which the orbital axis is oriented vertically and the rotational axis is tilted.

As the Earth orbits the Sun, the rotational axis remains pointed in the same direction in space. The Earth's North Pole (by convention, the one toward the top of each Earth) tips toward or away from the Sun as the Earth moves around its orbit. On the left Earth, the North Pole has its greatest tilt away from the Sun. This position is called the **winter solstice (WS)**. On the right Earth, the North Pole has its greatest tilt toward the Sun. We call this position the **summer solstice (SS)**. At the intermediate positions shown, the North Pole is tilted neither toward nor away from the Sun; these positions are called the **vernal equinox (VE)** and the **autumnal equinox (AE)**.

Dates

Because the Earth's revolution period is one year, the equinoxes and solstices occur on an annual basis at about the same time each year. The dates of each of these events are as follows:

Vernal equinox (or March equinox)	March 20–21
Summer solstice (or June solstice)	June 20–21
Autumnal equinox (or September equinox)	September 22–23
Winter solstice (or December solstice)	December 21–22

The main reason for the two-day spreads is that the length of the year is not an integral number of days. Each year contains a fractional day—every four years this fractional portion adds up to an extra day which is inserted during leap year.

Thus, while the vernal equinox appears to move around over a two-day interval from year to year, it is really our calendar that shifts to keep pace with the astronomical event. Additionally, the time and date depend on our longitude (east/west position on the Earth); the dates shown on the previous page are chosen to match the variation in solstice and equinox dates experienced by North American observers during the current era.

The Sidereal Year

How long is one year? As with the length of the day, it depends on how we define it. We can define the year in terms of the stars. To measure the **sidereal year**, line up the Earth, the Sun, and a star, start the revolution, and measure how long it takes to get the Earth, Sun, and star in line once again. The sidereal year is 365.256366 days, or $365^d6^h9^m10^s$.

The Tropical Year

To measure the **tropical year**, use the vernal equinox as a reference, measuring time from one vernal equinox to the next. The length of the tropical year is 365.242199 days, or $365^d5^h48^m46^s$, only slightly shorter than the sidereal year.

The tropical year is used for our calendar because it keeps better track of seasons than the sidereal year.

autumnal (or September) equinox
the point on the ecliptic where the Sun crosses the celestial equator from north to south; marks the start of autumn in the northern hemisphere

revolution period
the time it takes a body to complete one orbit

sidereal year
the interval between successive alignments of the Earth, the Sun, and a fixed star—about $365^d6^h9^m10^s$

summer (or June) solstice
the northernmost point on the ecliptic where the Sun is directly above the tropic of Cancer; marks the beginning of summer in the northern hemisphere

tropical year
the interval between successive alignments of the Earth, the Sun, and the vernal equinox—$365^d5^h48^m46^s$

tropic of Cancer
the latitude that marks the northern limit at which the Sun can be seen at the zenith, about $23\frac{1}{2}$°N; the Sun is overhead at the tropic of Cancer on the summer solstice

tropic of Capricorn
the latitude that marks the southern limit at which the Sun can be seen at the zenith, about $23\frac{1}{2}$°S; the Sun is overhead at the tropic of Capricorn on the winter solstice

vernal (or March) equinox
the point on the ecliptic where the Sun crosses the celestial equator from south to north; marks the start of spring in the northern hemisphere

winter (or December) solstice
the southernmost point on the ecliptic where the Sun is directly above the tropic of Capricorn; marks the start of winter in the northern hemisphere

Seasons

FIGURE 3.10A

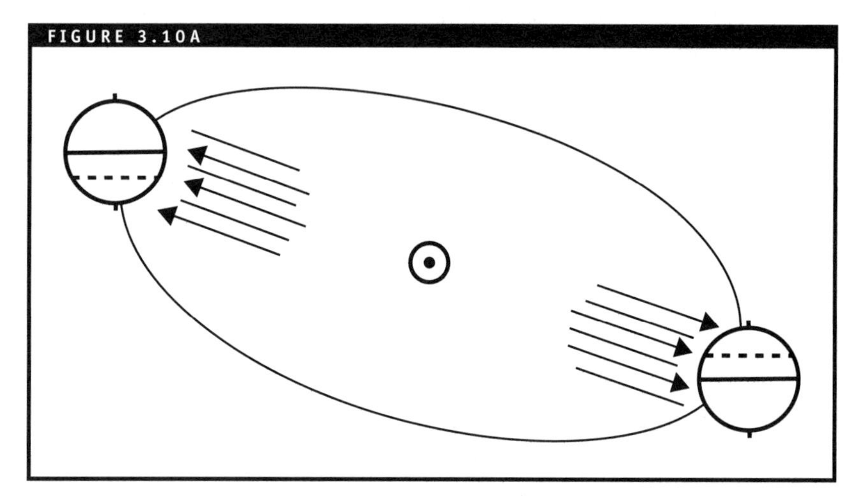

The solstices and equinoxes have seasonal-sounding names because there is a link between the motions of the Earth and the seasons. To see what causes seasons, examine the orientation of the Earth and Sun at specific points in the orbit—the winter and summer solstices.

Because the Earth is a sphere, the Sun's rays strike the surface at angles that vary in measure with their location on the Earth. The Sun's rays always fall most directly—from most nearly overhead—on the part of the Earth's surface that lies on a line between the centers of the Earth and Sun. The portion of the surface where this occurs varies as the Earth moves around in its orbit, as shown in Figure 3.10a.

FIGURE 3.10B

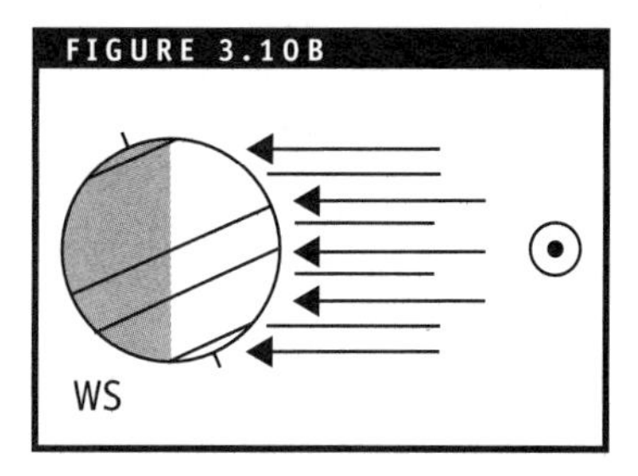

FIGURE 3.10C

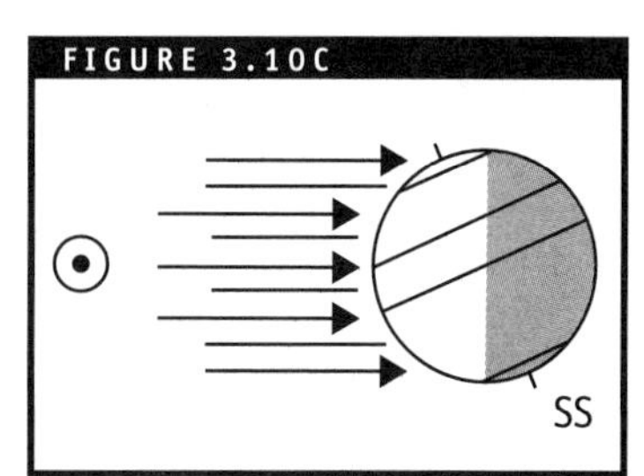

Figure 3.10b shows that during the time of the winter (or December) solstice, the location receiving the most direct rays lies on the tropic of Capricorn, at 23½ degrees south latitude. The heating effect of the Sun is greatest in the southern hemisphere, producing summer conditions there while winter occurs in the north.

Figure 3.10c shows that on the summer (or June) solstice, the reverse is true: Rays are most direct in the northern hemisphere, falling vertically on the tropic of Cancer. Greatest heating occurs in the northern hemisphere, producing summer there, while the southern hemisphere is having winter.

To the north or south of the direct rays, the heating effect is diminished because the Sun's rays are spread out over a greater portion of the surface, providing less energy per unit area. This can be seen by shining a flashlight at a flat surface from different angles (see Figure 3.11). When the flashlight beam shines straight down on the surface, the rays cover the smallest area and are most concentrated. As the beam intersects the surface at a shallower angle, the beam spreads out to cover a larger area. This effect can also be demonstrated by shining a flashlight on different parts of a globe and noting the area covered by the beam.

FIGURE 3.11

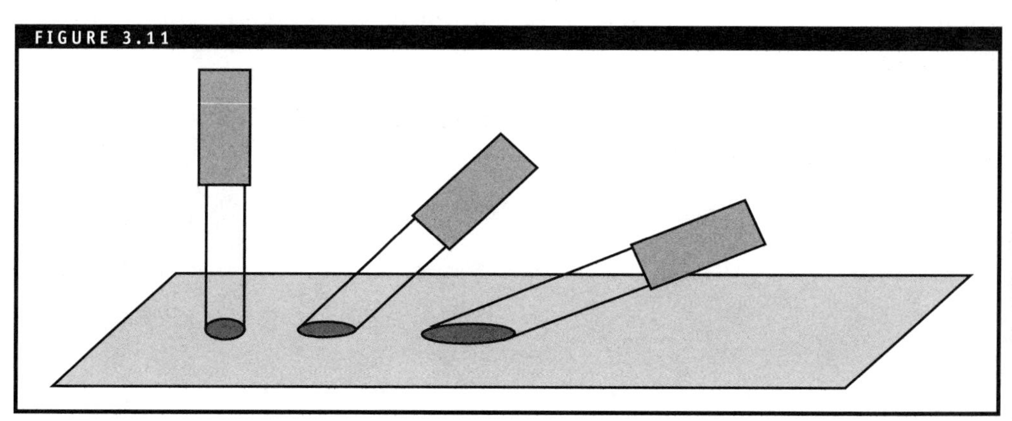

Day/Night Length

An additional effect that the tilt has on the seasons can be seen in Figure 3.12, which shows the variation in day and night for various latitudes. In this figure, the North Pole is tilted toward the Sun, and more than half of the northern hemisphere is illuminated. Consequently, persons living at northern latitudes would spend more time in the daylight than night, and they would receive more energy from the Sun. At the equator, day and night are equal in length, while in the southern hemisphere, the nights are longer than the days. The variation increases as one moves farther from the equator.

FIGURE 3.12

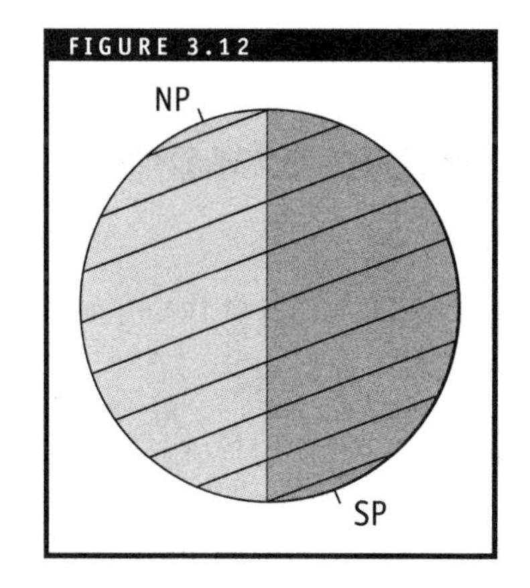

Seasonal Effects

The lengths of day and night also vary with the season. At the vernal (or March) equinox, the North Pole is tilted neither toward nor away from the Sun. As a result, all latitudes spend equal time in the daylight and darkness. As the Earth moves past the vernal equinox, the North Pole begins to tilt more and more toward the Sun, and daylight becomes longer in the northern hemisphere (and shorter in the south). The extreme is reached at the summer solstice, when the daylight is longest and the night is shortest for northern latitudes. Following the solstice, the days get shorter until they equal the nights at the autumnal (or September) equinox; the days continue to get shorter until the winter solstice, after which the trend again reverses. The southern hemisphere experiences the same changes but at opposite times—the days are longest at what is the winter solstice in the northern hemisphere, and shortest at what is the summer solstice in the northern hemisphere, with the equinoxes still having equal day and night.

Because the summer solstice occurs during winter in the southern hemisphere, there is some potential for confusion about the names of the solstices and equinoxes. This can be avoided by referring to the June solstice and the December solstice, the March equinox and the September equinox.

Effects of Latitude and Time of Year

The length of the days and nights depends on the latitude and the time of year. If you live at the equator, you have 12 hours of daylight and 12 hours of darkness each day of the year, because the Sun always illuminates one-half of the equator at a time. At noon, the Sun is straight overhead on the equinoxes (EQ), a point called the **zenith**. At the summer solstice, the noon Sun is north of the zenith, while at the winter solstice, it is south of the zenith. Thus, the noon Sun seems to oscillate back and forth, north and south of the zenith over the course of the year at the equator (see Figure 3.13a).

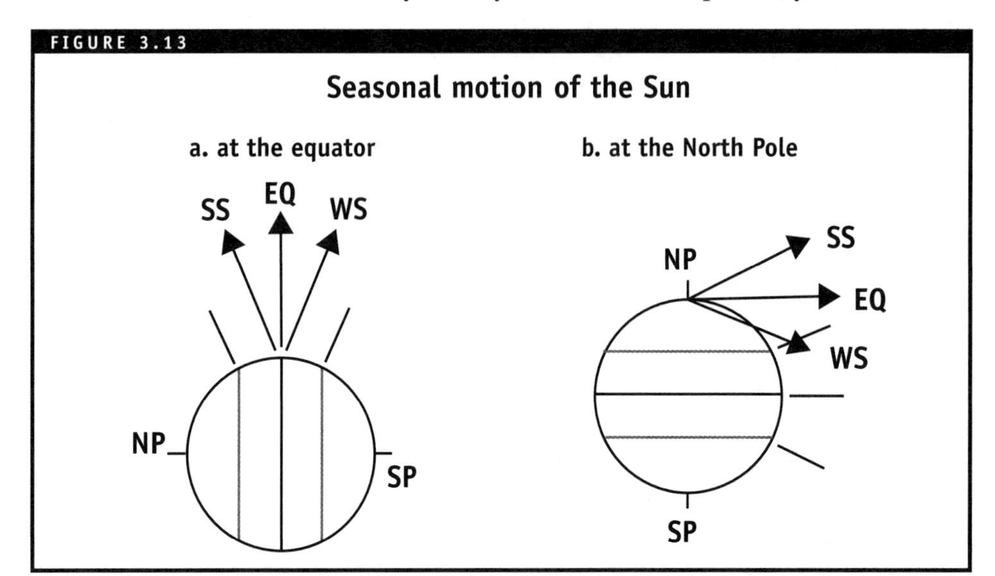

FIGURE 3.13
Seasonal motion of the Sun

THE MIDNIGHT SUN

At the North Pole, the story is different. At the equinox, when the Sun is overhead at the equator, it is on the **horizon**—the boundary between the sky and the ground—for an observer at the North Pole. At the summer solstice, when the Sun is north of the zenith at the equator, it appears north of the horizon at the North Pole, as shown in Figure 3.13b. As the Earth rotates, the Sun remains above the horizon, giving a day length of 24 hours, a phenomenon known as the **midnight sun**. At the winter solstice, the Sun remains below the horizon of a North Pole observer as the Earth rotates, resulting in a night that lasts 24 hours.

The Sun rises or sets when it crosses the horizon. For observers at the North Pole, the Sun is on the horizon only at the **equinoxes**. Thus, sunrise occurs at the vernal equinox and sunset happens at the autumnal equinox. This means that at the North Pole, there are six months of daylight following the vernal equinox and six months of darkness following the autumnal equinox.

FIGURE 3.12

NP

SP

THE ARCTIC AND ANTARCTIC CIRCLES

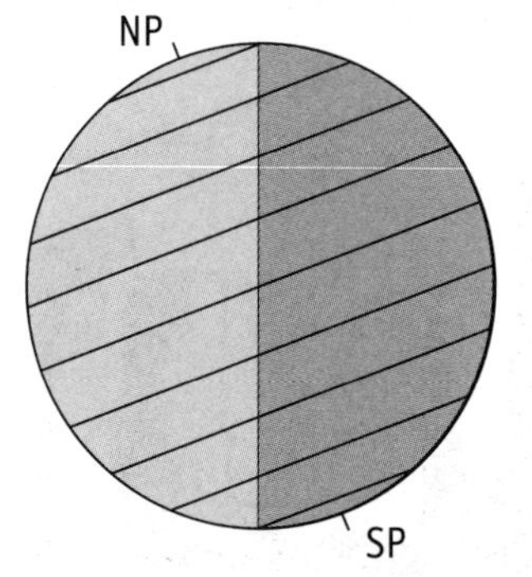

Clearly, the North Pole has a very strange day/night schedule, at least by our standards. Look again at Figure 3.12. When the Sun is to the left of the Earth (in the drawing), the North Pole is in sunlight. As the Earth spins, the pole remains in the sunlight, giving the midnight sun effect. A person located a few degrees of latitude away from the pole would also observe the midnight sun. In fact, you could move as far away from the pole as 23½° and still see the Sun at midnight, but only on the summer solstice. This position at 66½° north latitude is called the **Arctic Circle**; the corresponding location at 66½° south latitude is the **Antarctic Circle**. Persons located north of the Arctic Circle or south of the Antarctic Circle will experience at least some 24-hour days and an equal number of 24-hour nights during the year, with the number being greater for locations closer to the poles.

OTHER VARIATIONS

Between these two circles, the days and nights vary between 0 and 24 hours. At a middle latitude of 45°, the variation is from about 8½ hours to 15½ hours. The summer solstice has 15½ hours of day and 8½ hours of night, while the winter solstice has 8½ hours of day and 15½ hours of night. Closer to the equator, the variation is less extreme. The equinoxes have 12 hours of day and 12 hours of night.

At locations between the tropic of Cancer and the Arctic Circle, such as Minneapolis, Minnesota, at 45° north latitude, the noon Sun is found somewhere between the zenith and the observer's southern horizon, as shown in Figure 3.14. The Sun is highest in the sky (closest to the zenith) on the summer solstice, and lowest in the sky (closest to the horizon) on the winter solstice. The Sun reaches the zenith only in the tropics.

Seasons Revisited

The seasons—spring, summer, autumn, and winter—are caused by changes in the directness of the Sun's rays and the length of daylight at a given location: Direct rays and long days produce summer conditions. As for the duration of summer, any good calendar will tell you that the start of summer is the summer solstice, and its end is the autumnal equinox. Although the greatest heating effect, caused by the most direct rays and the longest daylight hours, occurs on the summer solstice, the hottest weather usually comes several weeks later, in late July or early August—the middle of summer. This delay is called **thermal lag**, or the cold pizza effect. When you pull last night's pizza out of the refrigerator to heat it for lunch, you turn on the oven to 300° and throw in the pizza. If you reach in and pull out the pizza one minute later, you will find it is still cold; in fact, the oven is not warmed up yet. If you give the oven time to warm up and then toss in the pizza for one minute, it will still be cold when you pull it out because it takes time to transfer enough heat to the pizza to raise its temperature to the desired level.

FIGURE 3.14

Seasonal motion of the Sun at 45° north latitude

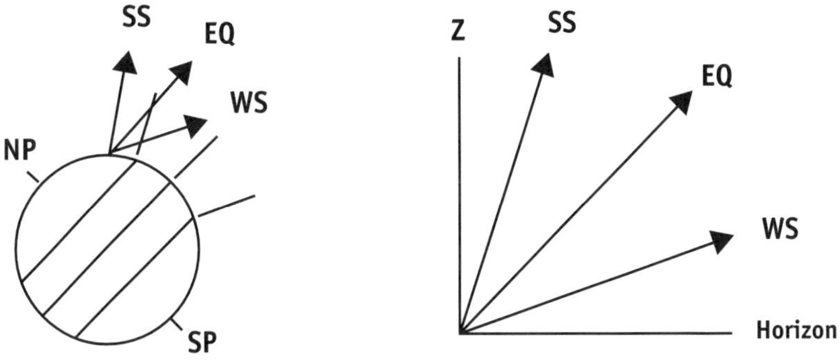

Antarctic Circle
66½° south latitude; the northernmost points in the southern hemisphere where the Sun can remain above (or below) the horizon for 24 hours at certain times

Arctic Circle
66½° north latitude; the southernmost points in the northern hemisphere where the Sun can remain above (or below) the horizon for 24 hours at certain times

equinox
the time when the Sun crosses the plane of the Earth's equator, making night and day of approximately equal length all over the world

horizon
the circle around you where the sky and the ground meet

midnight sun
the phenomenon of the Sun's remaining above the horizon at midnight (and throughout a 24-hour period); occurs only for observers north of the Arctic Circle or south of the Antarctic Circle

thermal lag
the time required for an object to heat up or cool down in response to changes in the rate of incoming heat

zenith
the point in the sky directly above the observer, opposite the nadir

FIGURE 3.15

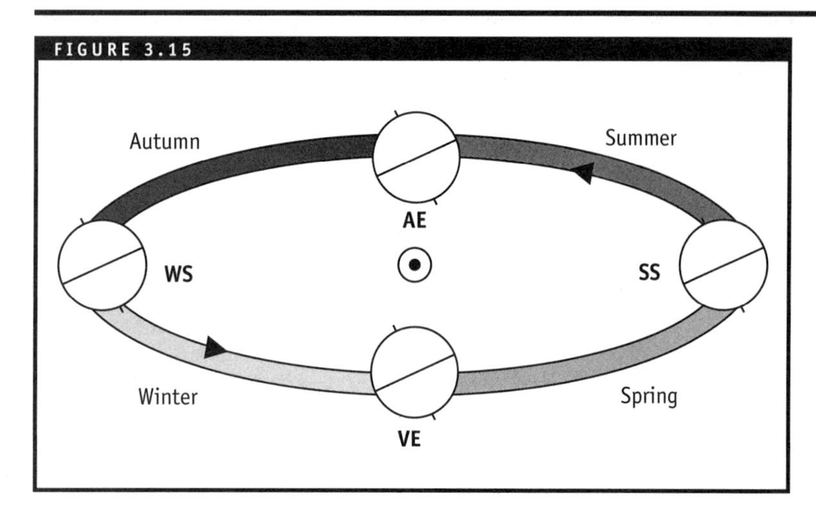

In the same fashion, the Sun transfers heat to the Earth. The northern hemisphere gets rather cold in the winter (due to indirect rays and short days), and it spends the spring warming up. By the summer solstice, after six months of lengthening days and increasingly more direct rays, the north is still warming. In fact, it does not reach its peak temperatures until midsummer, when the days have been getting shorter for several weeks. As a result of thermal lag, the seasons begin and end on the solstices and equinoxes, as shown in Figure 3.15.

Most of us think of summer as hot, winter as cold (or colder), and spring and autumn as in-between seasons that join summer and winter. We may perceive spring and autumn as similar to each other.

Figure 3.15 shows that although they are similar in temperature, they are quite different in other aspects. Astronomically speaking, spring is more like summer, while autumn is like winter. In both spring and summer the days are longer than the nights. For every hot summer day in late July there is a spring day in May when the Sun gets just as high in the sky, the rays are just as direct, and the day is just as long. The difference is due to the north or south motion of the Sun and thermal lag. The year is divided by the equinoxes: spring and summer are equivalent; autumn and winter are equivalent.

FIGURE 3.16

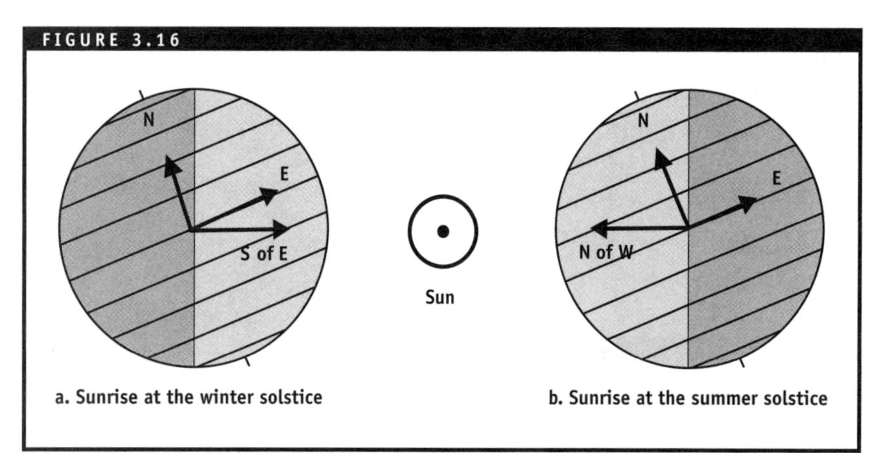

a. Sunrise at the winter solstice

b. Sunrise at the summer solstice

Sunrise and Sunset

The Sun appears to move slowly northward and southward in the sky over the course of a year. We can observe this motion, slow as it is, and safely measure the Sun's position.

The Sun is relatively safe to observe as it rises or sets because at those times of day its rays must pass through more of the Earth's atmosphere and become less intense. We can see evidence of its gradual movement northward or southward because the points on the horizon where the Sun rises and sets change with the seasons.

Figure 3.16a shows the Earth at the winter solstice position; the Sun illuminates half of the Earth, as shown. Because the Earth rotates eastward, the boundary shown between night and day marks the location where people are experiencing sunrise. If all these people point toward the Sun, they will point in the direction of the arrow labeled *S of E (south of east).* Figure 3.16b shows the Earth at the summer solstice position; the boundary shown is where the Sun is setting. For observers along this line, the Sun will set north of west, as indicated by the arrow.

Figure 3.17 shows the view from the opposite direction in space. From this vantage point we observe the sunrise boundary at the summer solstice (where the Sun rises north of east, as in Figure 3.17a) and the sunset boundary at the winter solstice (where the Sun sets south of west, as in Figure 3.17b).

FIGURE 3.17

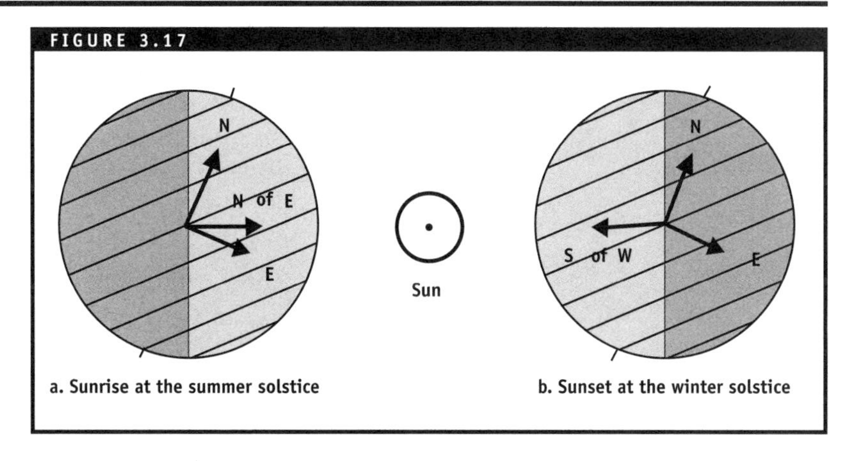

a. Sunrise at the summer solstice

b. Sunset at the winter solstice

The solstices mark the extreme directions of sunrise and sunset. During the rest of the year, the sunrise and sunset will vary: On the equinoxes—and only then—the Sun rises directly east and sets directly west; during autumn and winter, it rises south of east and sets south of west; during spring and summer, it rises north of east and sets north of west.

Shadows

Another way to follow the motion of the Sun is to observe shadows, noting both their directions and lengths. Since shadows are safer to look at than the Sun, they can be better indicators of the Sun's motion than the Sun itself.

A shadow always points away from the Sun. If the Sun is in the eastern sky, shadows point to the west; if the Sun is in the western sky, shadows point to the east. During the middle of the day when the Sun is in the southern sky, shadows point to the north. The exact directions will vary as the Sun moves across the sky. The length of a shadow depends on how high in the sky the Sun is—the higher the Sun, the shorter the shadow, as shown in Figure 3.18.

FIGURE 3.18

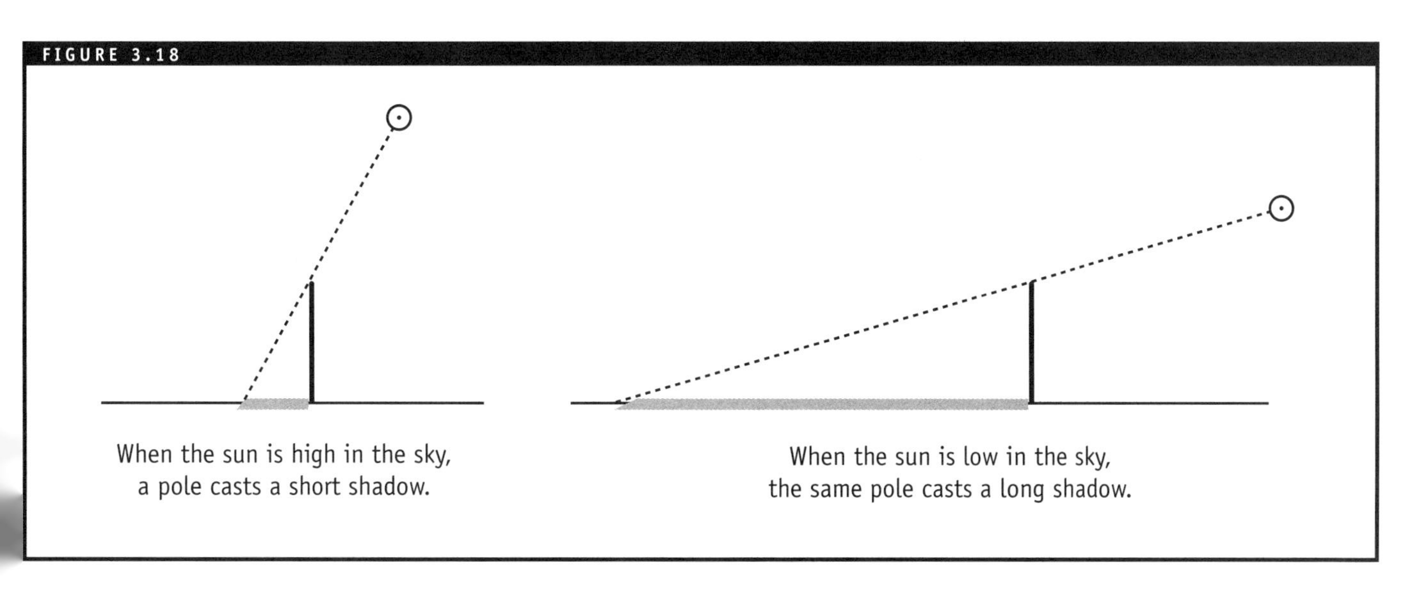

When the sun is high in the sky, a pole casts a short shadow.

When the sun is low in the sky, the same pole casts a long shadow.

Go outside on a clear day and find a vertical pole, such as a flagpole or a fence post, that is casting a shadow, or use a meter stick or a dowel and stand it upright. Find the tip of the shadow cast by the top of the pole and place a rock or other small marker at the tip. After a very short wait, you should notice the shadow tip moving away from the marker. If you continue placing markers at each new position of the shadow tip, you will trace its motion, and hence, the reverse of the Sun's motion. Over the course of the day, a generally (but not exclusively) eastward motion of the shadow tip will result from the Sun's westward drift. You should also see the shadow get shorter as the Sun moves higher in the morning and get longer as it moves lower in the afternoon.

The north-south motion of the Sun with the seasons can also be traced, but this motion takes longer to notice. This exercise also requires the shadow of some particular point on an object such as the top of a pole or a dot drawn on a window pane. Mark the location of the shadow at the same time each day, say 10:00 each morning. You should notice the shadow markers moving farther to the north during the summer and autumn (as the Sun moves toward the south) and then back to the south in the winter and spring following the winter solstice. Over the course of a whole year, your markers for this exercise will trace out a figure-8 shape called an **analemma.**

On the following pages you will find several activities to use for exploring shadows and the movement of the Sun.

analemma
the figure-8 shaped path traced over the course of a year by the Sun at a particular time of day

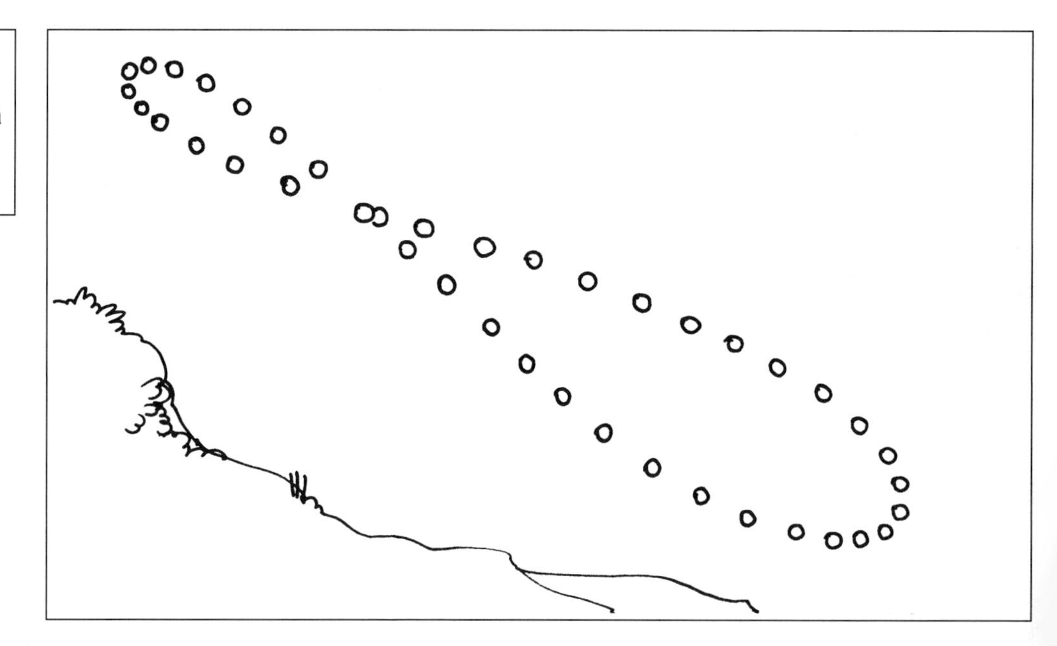

Shadow Experiments

Supplies you will need:

9" x 12" board

One-inch nail with a small head

Paper that covers the surface of the board

Tools you will need:

Hammer

Permanent marker

Construct a simple shadow-measuring device to follow two different motions of the Sun:

- The daily westward drift of the Sun across the sky
- The annual north-south migration of the Sun with the seasons

Steps to follow:

1. Draw an *X* on your board from corner to corner.
2. Tap the nail into the center of the *X*, which is the center of the board.
3. Label the board with the directions *North* (12" side), *East* (9" side), *South* (12" side), and *West* (9" side).
4. Slip paper over nail. Label paper with *N, E, S,* and *W*. You can use different papers for different observations.

Observations

I. Orient the shadow board in the sunshine and mark the position of the shadow of the nail's head at regular intervals during the day. Use the same orientation for the board each time.

II. Repeat the experiment the next day.

III. Repeat the experiment the next week.

Results of observations should be similar to these: The shadow becomes shorter toward midday, then lengthens in the afternoon. Local noon will occur when the shadow points north; this will probably not occur at noon by the clock. In spring and summer, a line connecting the shadow tips should be concave to the south; in fall and winter, it should be concave toward the north.

Repeating the experiment the next day should produce essentially the same results. However, a repeat made after a week or two should show subtle changes in the shadow lengths and directions for corresponding times, due to the northward or southward seasonal motion of the Sun.

The annual motion can best be shown by marking one shadow tip at the same time of day, on one day each week over the course of the year. Use the same sheet of paper for the whole year. As the Sun moves south for the winter, the shadow tip will move north. The maximum shadow length should occur on the winter solstice, after which the Sun will head back to the north. To generate the complete path of the shadow tip (a figure-8 shape called an analemma), this project will have to be undertaken for the entire year.

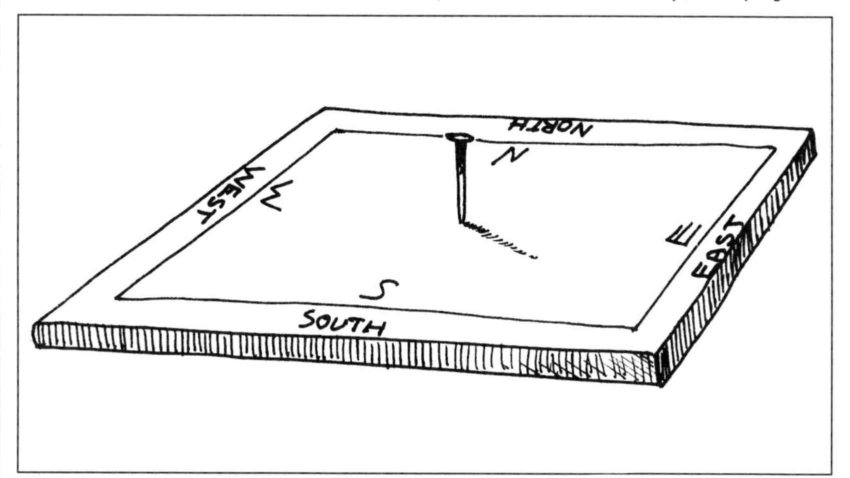

Note to teachers: This project is a great classroom experiment. Students make predictions, build appropriate testing equipment to test their predictions, make measurements, interpret the results, and analyze the whole process for errors. It can be adapted for several grade levels by emphasizing different aspects of the scientific method.

Making and Using a Sundial

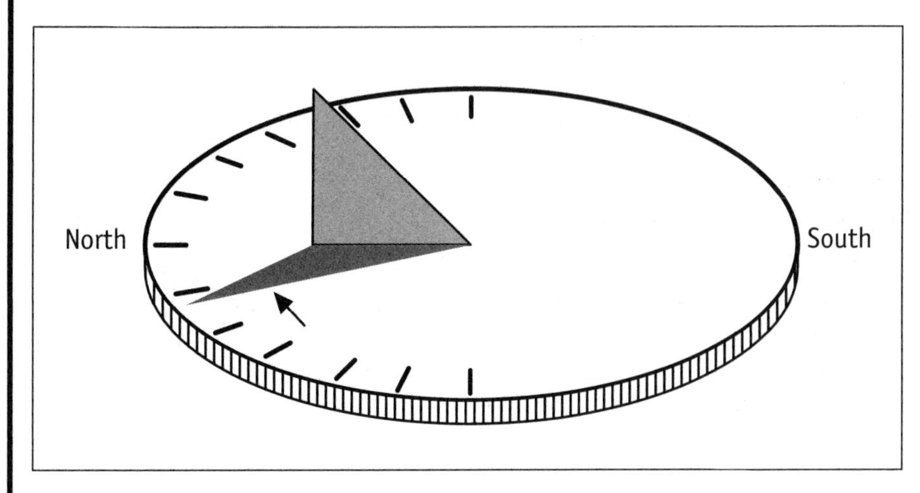

A sundial indicates the time of day by measuring the position of the shadow of a gnomon on the dial. Since the length and direction of the shadow of a *vertical* gnomon would vary with the season, the gnomon must be inclined from the vertical to be parallel with the Earth's rotational axis. In the northern hemisphere, this requires that the gnomon point northward and make an angle with the horizontal, which is equal to the observer's latitude.

The shadow of the gnomon's sloping edge (arrow) is used to mark the time. This shadow indicates the time is about 10:45 a.m. (see illustration below).

The positions of the hour lines on the dial depend on the latitude at which the sundial is used. Angles required for a latitude of 45° are shown in the diagram.

Even using the proper angles, a sundial still does not keep clock time because the rate of the Sun's apparent motion through the sky varies slightly through the year. In addition, the sundial measures local time, while your clock gives the time at the central meridian of your time zone. Therefore, your sundial will probably show variations of up to 20 minutes from the time on your watch.

For more information, see *Sundials: Their Theory and Construction* by Albert Waugh (Dover Publications, 1973).

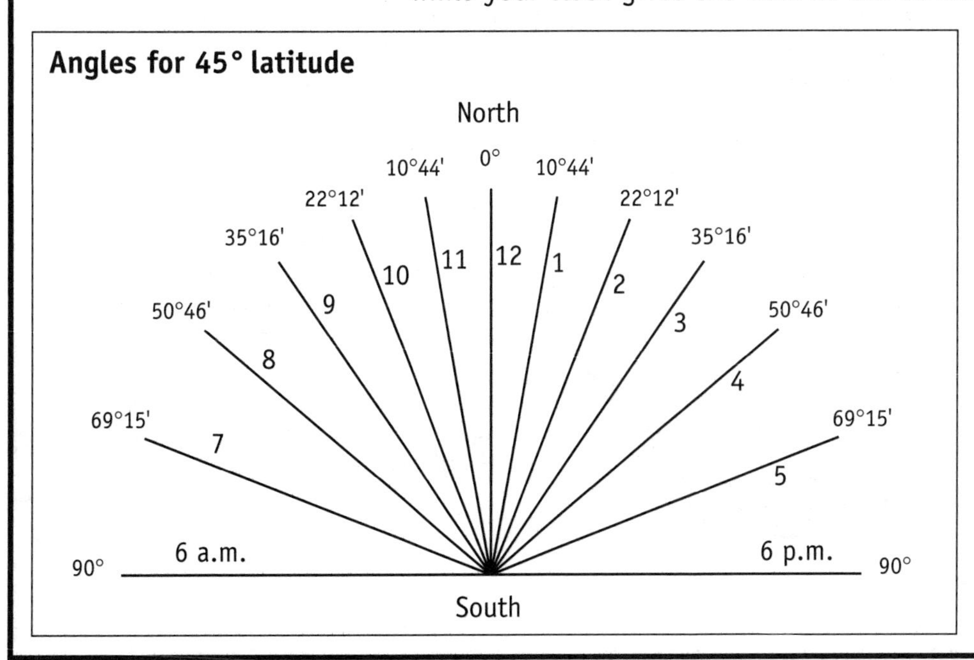

Sundial Face

Find out the latitude at your location and trace the appropriate lines on a separate sheet of paper to create an accurate sundial face. Refer to either of the following Internet sites for information about how to make a sundial: *http://www.sundials.co.uk* and *http://www-spot.gsfc.nasa.gov/stargaze/Sundial.htm.*

Latitude	Noon	1 p.m./11 a.m.	2 p.m./10 a.m.	3 p.m./9 a.m.	4 p.m./8 a.m.	5 p.m./7 a.m.	6 p.m./6 a.m.
- - - - 45°	0°00	10°44'	22°12'	35°16'	50°46'	69°15'	90°
.... 40°	0°00	9°46'	20°22'	32°44'	48°04'	67°22'	90°
-·-·-·- 35°	0°00	8°44'	18°29'	29°50'	44°49'	64°58'	90°
——— 30°	0°00	7°38'	16°06'	26°34'	40°54'	61°49'	90°

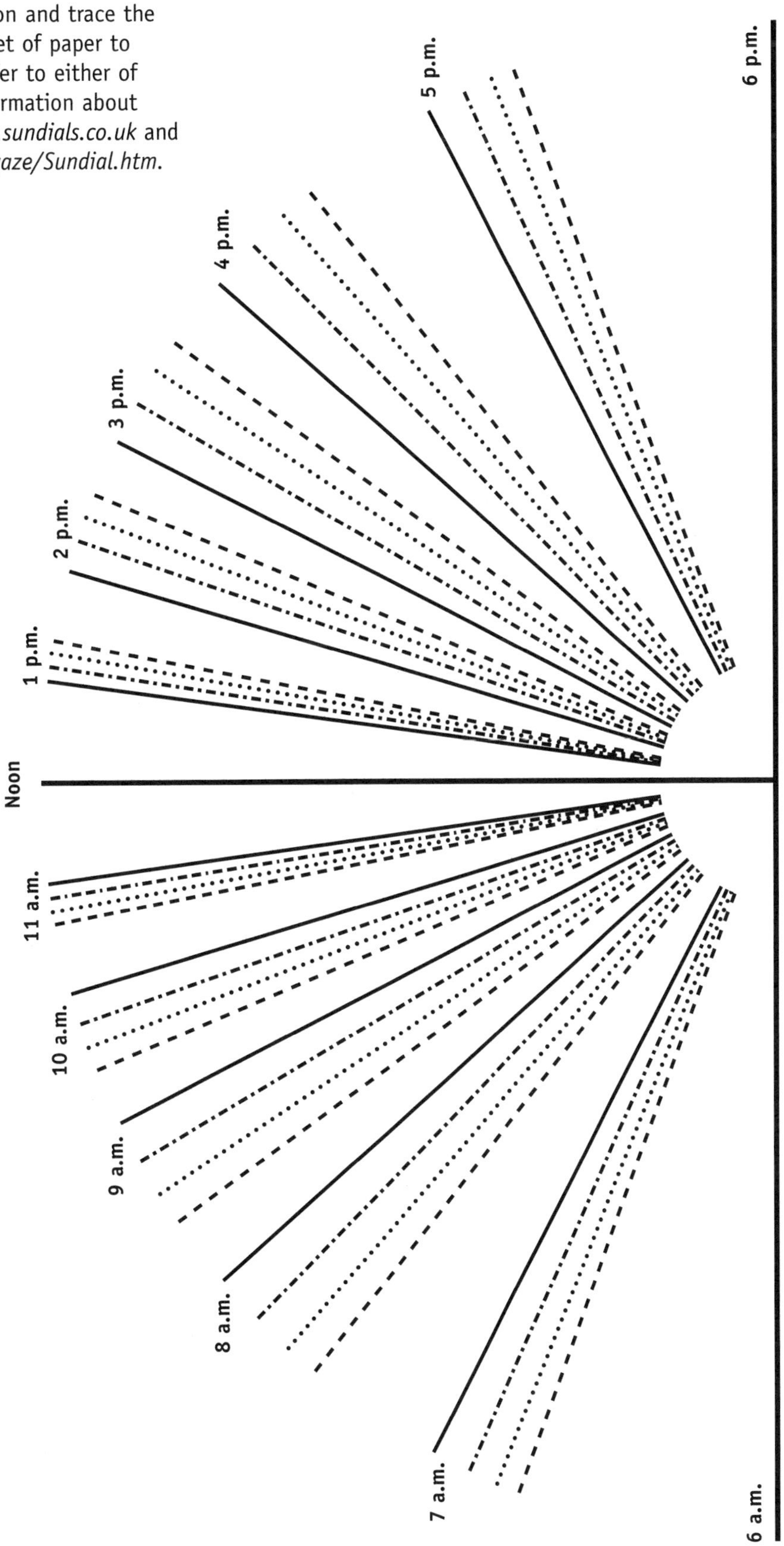

6 p.m.
5 p.m.
4 p.m.
3 p.m.
2 p.m.
1 p.m.
Noon
11 a.m.
10 a.m.
9 a.m.
8 a.m.
7 a.m.
6 a.m.

Interactive Model: Diurnal Motion of Astronomical Bodies

Make and use this model to illustrate and explore the motion of objects in the sky caused by the rotation of the Earth on its axis.

Materials needed:

Circle of ⅜-inch plywood with a diameter of 11"

Compass (the kind used to draw circles)

Small block of wood the length of the compass point

5¼-inch-long, 5/16-inch dowel with a 1/16-inch diameter hole in one end (parallel to the dowel's axis)

12" bicycle wheel spoke

½-inch-diameter rubber ball

Protractor

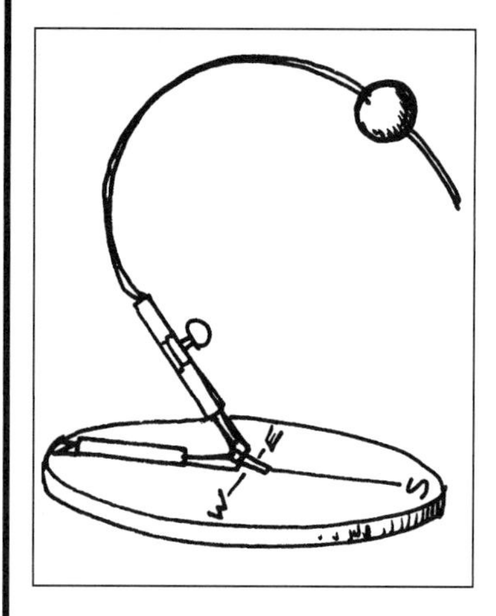

To make the model:

1. On the plywood circle, mark the directions *North, South, East,* and *West* as shown in the diagram.
2. Mount the compass on the plywood circle. Use the small wooden block to affix the compass to the *North* direction line. Position the compass so that when the pencil arm is raised perpendicular to the board, the pencil will be directly over the center point of the plywood circle.
3. Replace the compass pencil with the dowel.
4. Spear the rubber ball through the middle with the bicycle spoke.
5. Screw the threaded end of the spoke into the hole in the dowel, and bend the spoke to follow the circular arc of the plywood.

To use the model:

The plywood base represents the horizon plane of the observer, who is located at the center of this base where the East-West line intersects the North-South line. The rubber ball represents an object (the Sun, the Moon, star, and so on) on the celestial sphere, a portion of which is represented by the bicycle spoke. The wooden dowel represents the axis around which the celestial sphere rotates; this axis will be set parallel to the rotational axis of the Earth by adjusting the angle of the compass.

- To set the desired latitude—adjust the compass angle by measuring with the protractor.
  - 0° (horizontal) orients the model to what can be seen at the equator.
  - 90° (vertical) orients the model to what can be seen at the North Pole.

  Angles between these place the observer at different latitudes: e.g., 45° for Minneapolis; 40 for Philadelphia, Denver, Salt Lake City; 35° for Chattanooga, Albuquerque; 30° for Houston, New Orleans, Jacksonville, and so on. (Placing the center of the model tangent to the desired location on a globe may help clarify why this works—the dowel should be parallel to the axis of the globe in each case.)
- Set the angle to 0°, move the spoke to the eastern horizon, and slide the ball to the east point on the horizon. The ball now marks the location of the Sun at both the vernal and autumnal equinoxes. Rotation of the spoke makes the ball rise higher, which simulates sunrise at the equator. Continued motion of the wire will show that after rising due east, the Sun passes through the zenith at noon and sets due west.
  - Changing the compass angle will show how the Sun moves on the vernal equinox as seen from other latitudes. Note that the Sun continues to rise due east and set due west, but no longer passes through the zenith.
  - Changing the position of the ball on the spoke will show how the Sun moves on other days of the year. Move the Sun south along the wire for autumn and winter, or north for spring and summer. By sliding the ball to 23 ½° north of east (the summer solstice) and setting the latitude to 90° (the North Pole), the midnight sun phenomenon can be demonstrated—the Sun will circle the horizon without setting.
  - To view star trails, look across the model from the observer's position in the center, where the spoke is. Place the ball along a portion of the spoke that is visible and rotate the spoke as before. When the angle is set for mid-latitudes; the ball should slant up to the right when looking east. It should arc downward when looking south and upward when looking north along the horizon. Circumpolar stars can be simulated by moving the ball close to the North Celestial Pole position and rotating the spoke. (See Chapter 8 for information about circumpolar stars and the celestial poles.)
  - Another way to explore this motion is to hold the spoke still and rotate the plywood base. This demonstrates that the same motion will be seen by a rotating observer inside a stationary celestial sphere.

Chapter 4
Moon Motions and Phases

Revolution

The Moon has motions similar to those of the Earth. The Moon revolves around the Earth in a counterclockwise direction, much as does the Earth around the Sun. In comparing the relative distances of the Sun and Moon, the Sun-Earth distance is about 400 times the Earth-Moon distance. The Moon's orbit around the Earth is considerably smaller than the Earth's orbit around the Sun, and the Moon takes less time to make one orbit than does the Earth. Chapter 3 demonstrated that you can measure the sidereal year by observing the revolution of the Earth with respect to a star. If you use the same basis to measure the Moon's revolution, you get the sidereal month, as shown in Figure 4.1.

The length of a sidereal month is about 27⅓ days. This means that every 27⅓ days, the Moon passes close to the same star, as seen from Earth. Its counterclockwise revolution carries it eastward among the stars (from right to left for a northern hemisphere observer facing south).

FIGURE 4.1

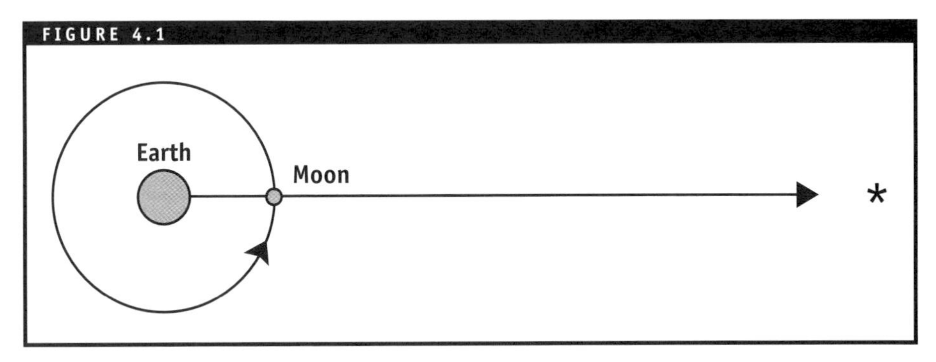

relative mass
Earth 1
Moon 0.0123
Sun 333,000

distance from Earth
Moon ≈ 384,000 km
Sun ≈ 150,000,000 km
≈ means approximately

Newton's law of gravity
For every two masses there is an attractive gravitational force, with each mass pulling on the other.

gravitational force equation

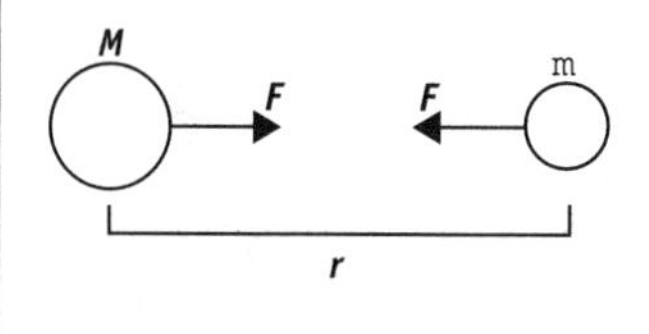

$$F = \frac{GMm}{r^2}$$

F = force

G = the gravitational constant

M_e = mass of the Earth

M_m = mass of the Moon

M_s = mass of the Sun

r_e = distance between Moon and Earth

r_s = distance between Moon and Sun

The force of the Sun on the Moon
$F_s = (G \times M_s \times M_m) \div r_s^2$

The force of the Earth on the Moon
$F_e = (G \times M_e \times M_m) \div r_e^2$

The ratio of the two forces
$F_s \div F_e = (M_s \div M_e) \times (r_e \div r_s)^2$
(G and M_m divide out)

The Sun's mass is 333,000 times the Earth's mass. The Sun-Moon distance is about 400 times the Earth-Moon distance.

$F_s \div F_e = 333{,}000 \div 400^2$
$= 333{,}000 \div 160{,}000$
$= \text{about } 2$

The Sun and Earth both act gravitationally on the Moon. By using Newton's gravitational force equation, calculate the force of the Earth on the Moon, and also the force of the Sun on the Moon. The Earth is closer to the Moon than the Sun is, but the Sun is much more massive than Earth. The result of the gravitational force calculation is that the Sun exerts a gravitational force on the Moon that is approximately twice as strong as the Earth's gravitational force on the Moon, implying that the Moon should be orbiting the Sun instead of the Earth! (See sidebar on this page.)

When you combine the Moon's revolution around the Earth with the Earth's orbit around the Sun, you might imagine that the Moon follows a looping path, like that shown in Figure 4.2. But each loop should represent a month, and there should be only 12 or 13 of these loops each year. Clearly, this diagram would have too many loops around the orbit to match reality.

A more accurate path is shown in Figure 4.3. In this picture, the Moon does not loop around the Earth at all; it continuously moves forward along the Earth's orbit, without backing up.

To understand how the Moon appears to circle Earth each month, imagine yourself driving a car along the freeway. When you approach a slower car in your lane, you pull out to pass in the left lane. As you pass the car, it appears to move backward on your right—an illusion created by your higher speed. Once you get ahead of the car, you pull back into the right lane; immediately, the other car pulls into the left lane and speeds up to pass you. You soon see the other car moving forward on your left and pulling in front of you. From your point of view, the other car has just completed an orbit around you. If you pull out to pass again and repeat the sequence of events just described, you will perform the same "leapfrogging" motion when the Earth and Moon when they move around the Sun together. Figure 4.4 shows that both the Moon and the Earth orbit the Sun, remaining reasonably close together.

FIGURE 4.2

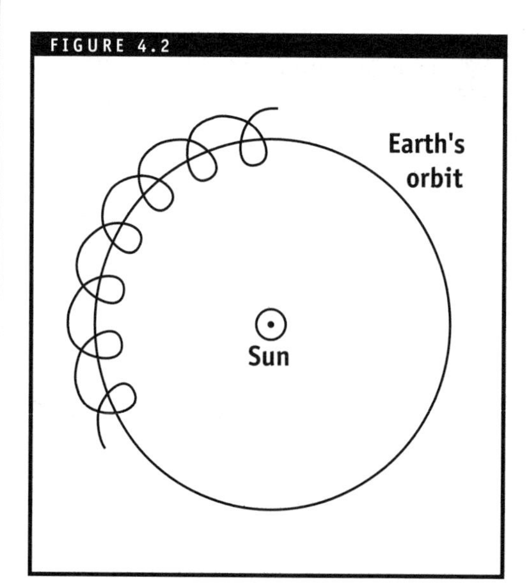

FIGURE 4.3

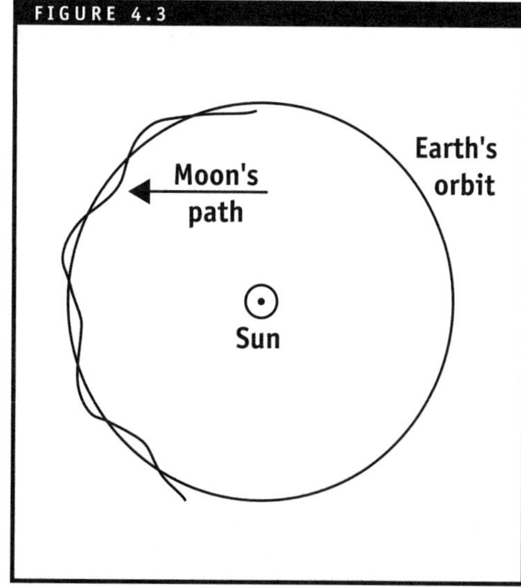

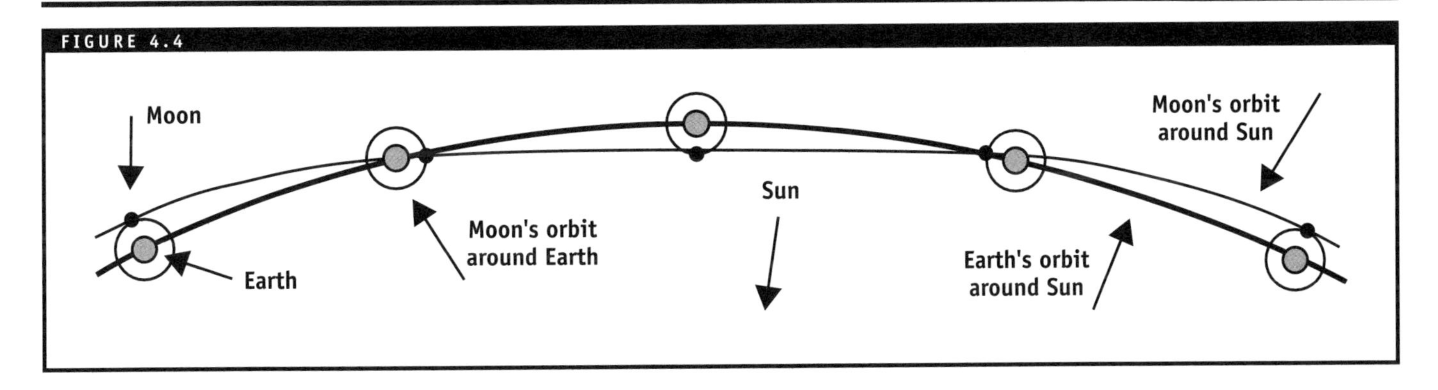
FIGURE 4.4

Rotation

The Moon revolves as the Earth does. An observer in space would see the Earth turning slowly on its axis. When you look at the Moon from the Earth, you can clearly see features on its surface that will easily show any similar motion. However, when you observe the Moon over a long period of time, you find that it always presents the same features and gives no hint of rotation. Does this mean that the Moon does not rotate?

Figure 4.5 shows the Moon at four different positions in its orbit around the Earth. On each figure of the Moon a line has been drawn that points toward the Earth. Earthbound observers looking at the Moon will always see the side with the line, no matter where the Moon is in its orbit.

But from your vantage point above the page, you can see that the line is not always pointing in the same direction—on the top Moon, it points down; on the left Moon, it points right; and so on. In order for the line to point in different directions (but always toward Earth), the Moon must be rotating. Furthermore, it must be rotating in the same direction that it revolves (counterclockwise) and at the same rate. That is, the Moon's sidereal rotation period is one sidereal month.

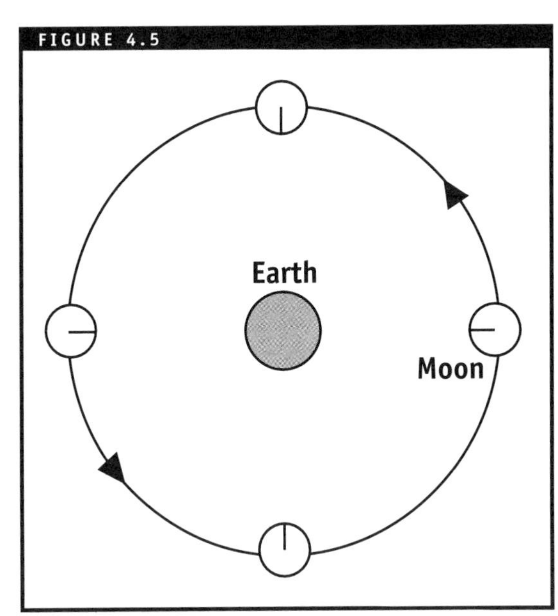
FIGURE 4.5

In astronomy, the apparent motion of an object often depends on the location and motion of the observer. In this case, observers on Earth might say that the Moon does not rotate, while those located on distant stars would contend that it does. Because the Earth itself is moving, and its motion affects our perception of other bodies' motions, astronomers generally choose the distant (and essentially motionless) stars as the standard reference. Rotation and revolution periods given are usually sidereal periods, unless otherwise stated.

Phases

Anyone who has observed the Moon has noticed that its shape changes gradually from day to day; these shapes are called the phases of the Moon. Of course, the Moon itself is not changing shape—it is our vantage point that changes. Figure 4.6 shows the basic geometry that produces phases.

The Earth and the Moon do not generate their own light; they both reflect light from the Sun. Because the Moon and Earth are spheres, the Sun illuminates only half of each at a time. Because the Moon is a sphere, we can only view half of it at a time—the half turned toward the Earth (the half on the lower left side of Line AB). Because the dark portion of the Moon is difficult to observe, we generally only see the portion of the Earth-facing half that is illuminated. This portion changes as the Moon moves around its orbit, giving us the different lunar phases.

FIGURE 4.6

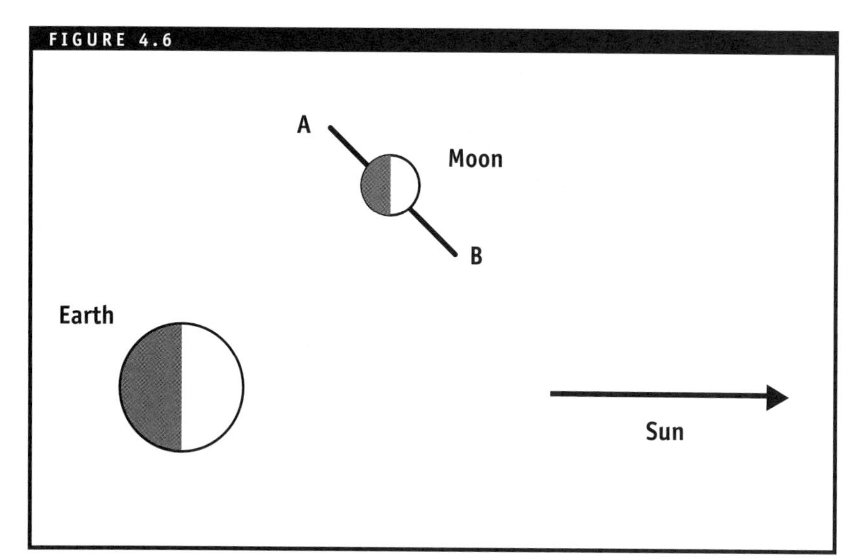

To see the range of phases, get a table tennis ball and paint half of it black. Then view the painted ball from all possible angles—the portion of white that you see will correspond to the Moon's phases. Whatever the phase, the line between the white and black regions of the ball always starts and finishes on opposite ends of a diameter.

Moon Phases as Viewed from Earth

The Moon's phases appear as the Moon moves in its orbit around the Earth. Each day, the Moon-Earth-Sun angle changes slightly, and a different portion of the Moon is illuminated. Figure 4.8 shows the Moon at eight different positions in its orbit around the Earth. An observer on Earth would see the Moon in the direction of the numbered arrow pointing toward it.

FIGURE 4.7

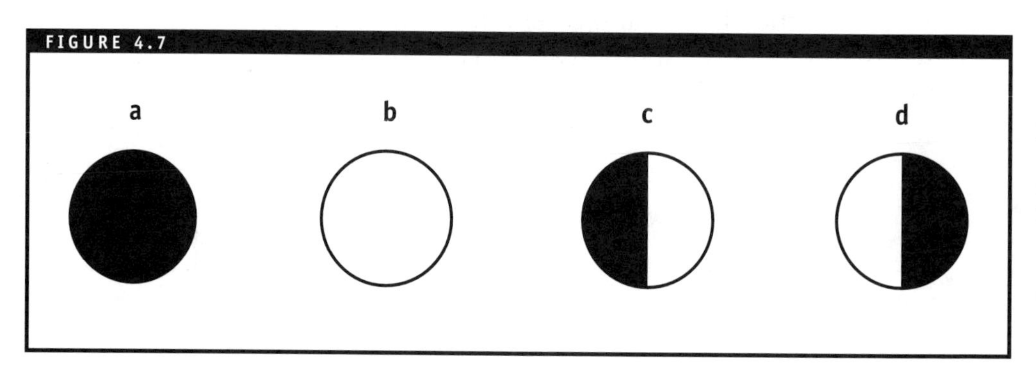

Some phases are fairly easy to understand. For example, an observer on Earth looking at Moon 1 would see only the dark side of the Moon, as shown in Figure 4.7a. Moon 5 would appear completely illuminated, as in Figure 4.7b. To see this, use Figure 4.8 on this page. Orient and rotate the page so that you are looking along the appropriate arrows from the Earth toward each Moon. Moon 3 will be half illuminated, with the bright side on the right (Figure 4.7c), while Moon 7 should also be half illuminated, but with the bright side on the left (Figure 4.7d).

These four phases are the principal phases of the Moon, as noted on calendars and in almanacs. They occur when the Moon is in line with the Earth and Sun or at right angles to the Earth-Sun line. As such, they occur at precise times, on certain days of the lunar cycle.

FIGURE 4.8

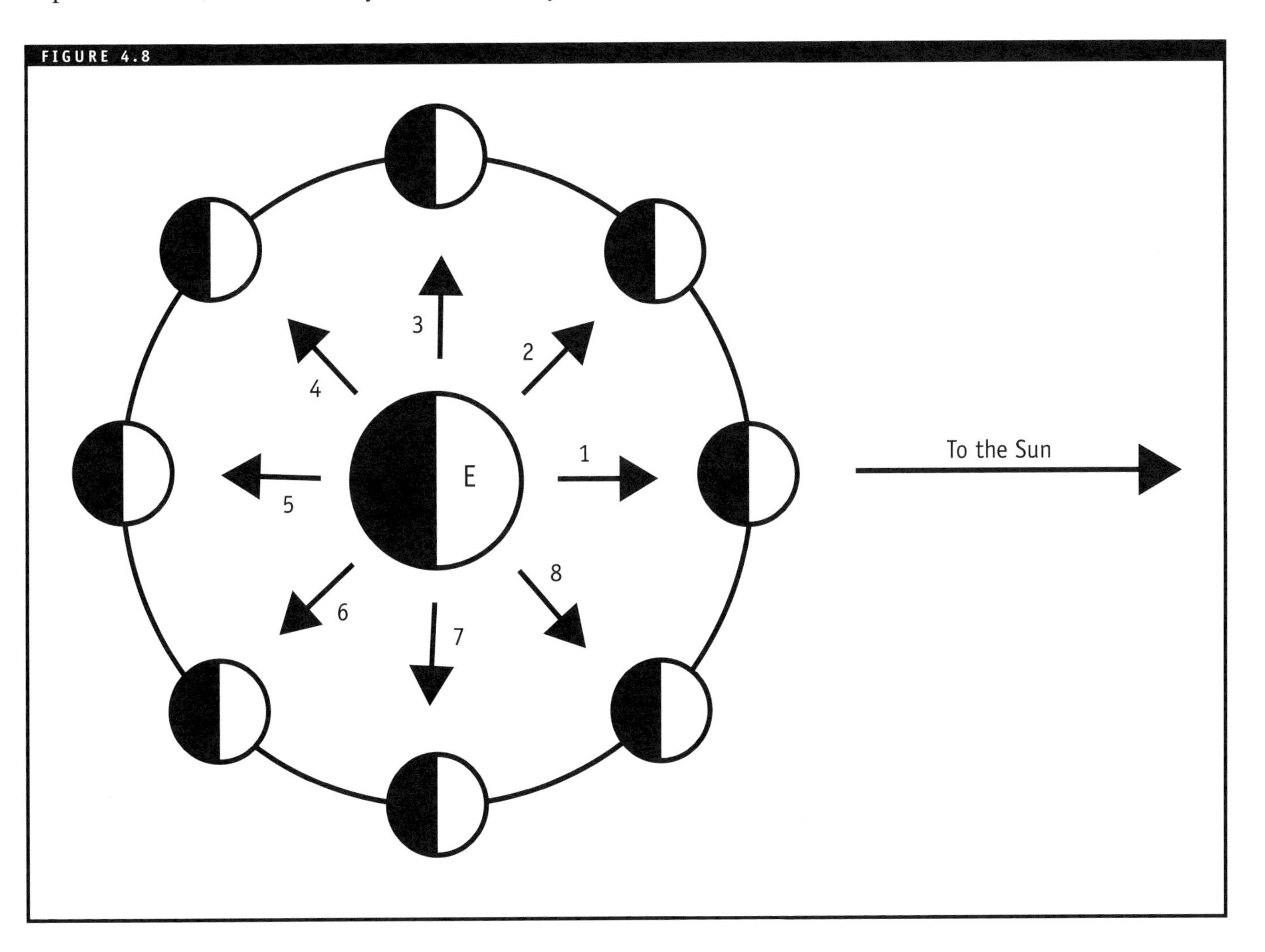

FIGURE 4.10

Phases of the Moon

New

Waxing crescent

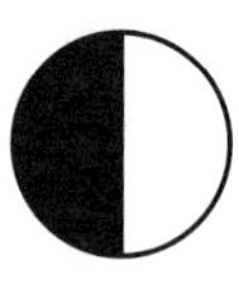

First quarter

Waxing gibbous

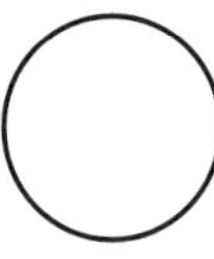

Full

Waning gibbous

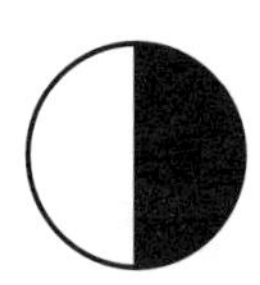

Last quarter

Waning crescent

FIGURE 4.9

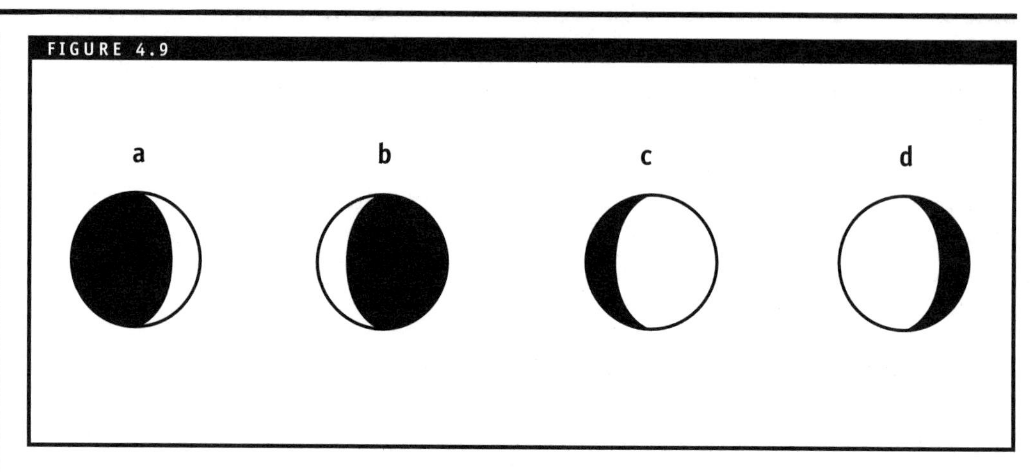

The remaining four phases are not as well known, nor are they as precisely defined. In fact, their shapes vary with time, with the amount of illumination increasing or decreasing, depending on the phase.

For example, referring again to Figure 4.8, Moon 2, shown midway between Moons 1 and 3, could occur anywhere between those two phases. As seen from Earth, Moon 2 would appear to be mostly dark, with the bright side on the observer's right. It should look something like Figure 4.9a. Moon 7 is similar, but with the bright side on the left, as in Figure 4.9b. Moon 4 will be mostly light, with the dark side on the left (Figure 4.9c), and Moon 6 will be the same, but with the dark side on the right (Figure 4.9d).

Note that the Moon does not always appear as shown in books—including this one—"standing" straight up and down, with light and dark portions on either the right or left. The curvature of the Earth and the observer's location might make the Moon appear to be on its side or even upside down compared to its depiction here. In fact, readers in the southern hemisphere would do well to rotate the book 180° when looking at the Moon phase pictures.

As the Moon moves counterclockwise around its orbit, it proceeds through all eight phases in numerical order, as shown in Figure 4.10. Also shown in this figure are the official names of the eight phases.

A new moon is all dark; a full moon is all light. The first quarter and last quarter moons are half illuminated and are sometimes called "half moons"—an unofficial term. The last quarter moon is also called a third quarter moon; either term is acceptable. A **crescent moon's** surface is *less* than half illuminated; a **gibbous moon's** surface is *more* than half illuminated.

When the Moon is **waxing** (from new to first quarter to full), the illuminated fraction of the face increases; when the Moon is **waning** (from full to last quarter to new), the illuminated fraction decreases.

The gibbous and crescent phases are most commonly observed—they fill in the time between the special alignments that cause the new, full, and quarter phases.

The Synodic Month

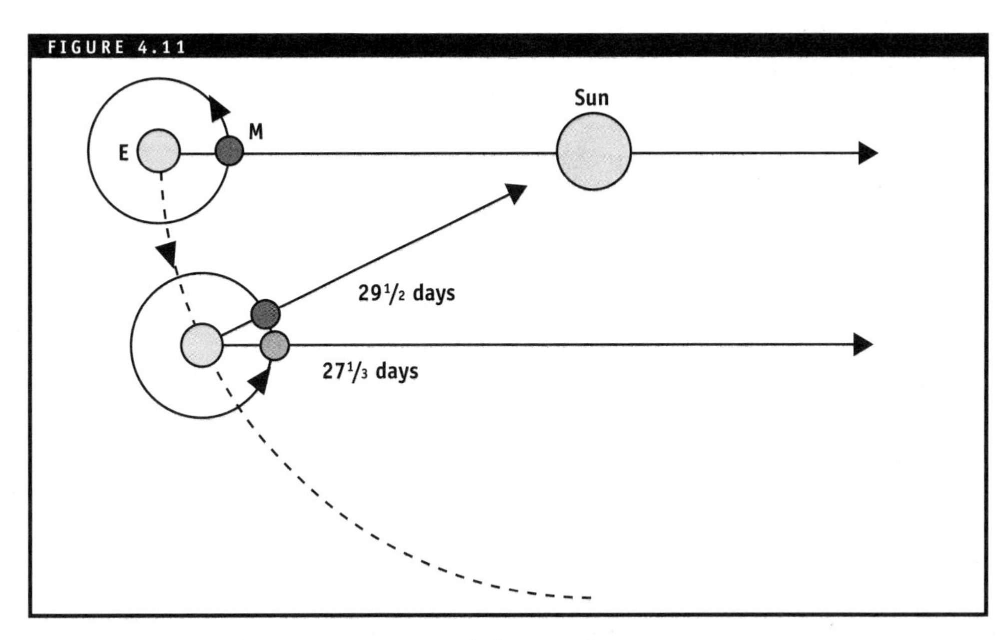

The Moon cycles through its phases in about 29½ days from new moon to new moon, an interval called a **synodic month**. This period is slightly different from the previously defined sidereal month, which is about 27⅓ days. The reason for the difference is that the months measure different periods—the **sidereal month** is the period of revolution of the Moon around the Earth with respect to the stars, and the synodic month is the period of revolution of the Moon around the Earth with respect to the Sun, which illuminates the Moon and causes its phases.

If the Earth did not also move during this time, the sidereal and synodic months would be the same. Because the Earth moves in its orbit around the Sun, the Moon needs about two extra days to catch up to the Sun after it has realigned itself with the distant star, as shown in Figure 4.11.

It takes the Moon about 29½ days to go from new moon to new moon or from full moon to full moon. The interval between new and full moons is about two weeks, while from new moon to first quarter is about one week. Thus, the principal phases occur at intervals of about one week, with the intervening crescent and gibbous phases each lasting about one week. Although the Moon may appear to be full for several days, it is really only full for an instant on one day. Before that time, it is a waxing gibbous, and afterward, it is a waning gibbous.

crescent moon
a phase between new and quarter that shows a surface that is less than half illuminated

gibbous moon
a phase between quarter and full that shows a surface that is more than half illuminated

sidereal month
the interval between successive alignments of the Earth, the Moon, and a fixed star—about 27.32 days

synodic month
measures the time required for the Moon to orbit once and line up again with the Earth and the Sun—about 29.53 days

waning moon
any phase during which the illuminated fraction of the Moon's face decreases (from full to last quarter to new)

waxing moon
any phase during with the illuminated fraction of the face increases (from new to first quarter to full)

Moonrise and Moonset

We have seen how the rotation of the Earth causes the Sun to rise and set (at most latitudes). The Earth's rotation also causes the Moon to rise and set. However, the constant orbital motion of the Moon makes it rise and set at times that vary from day to day as the Moon shifts its phases. In order to know when to watch for the Moon to rise, we must know its current phase.

The times of moonrise and moonset are easy to learn. Consider only the four principal phases, which rise and set at only four different times of day and night—sunrise, noon, sunset, and midnight.

earthshine
sunlight reflected from Earth that illuminates the dark face of a young crescent moon

Phases of the Moon

new
rises at sunrise; sets at sunset

waxing crescent
rises between sunrise and noon; sets between sunset and midnight

first quarter
rises at noon; sets at midnight

waxing gibbous
rises between noon and sunset; sets between sunrise and noon

full
rises at sunset; sets at sunrise

waning gibbous
rises between sunset and midnight; sets between sunrise and noon

last quarter
rises at midnight; sets at noon

waning crescent
rises between midnight and sunrise; sets between noon and sunset

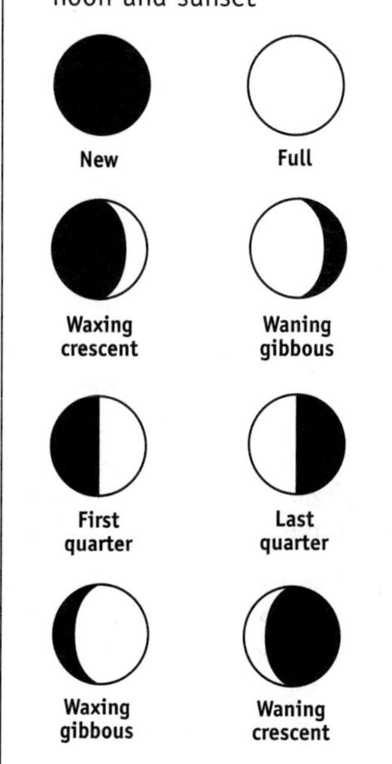

FIGURE 4.12

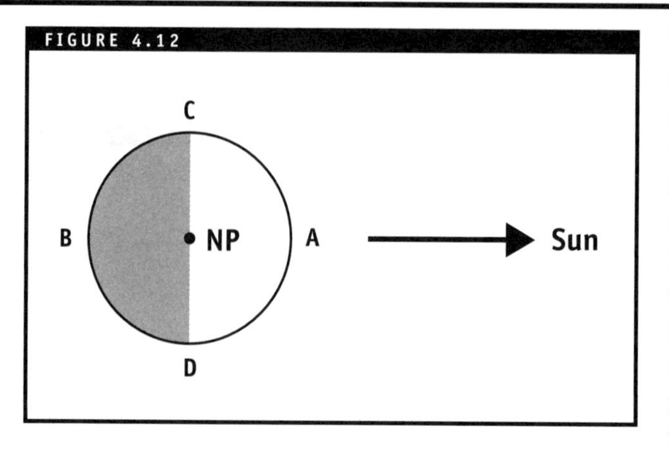

The time of day or night depends on your location on the Earth with respect to the Sun. For a person standing in the middle of the illuminated side of the Earth (Position A in Figure 4.12), the time should be noon, while midnight occurs for people in the middle of the dark side of the Earth (Position B). Observers at Positions C and D, on the boundary between light and dark, will be experiencing either sunrise or sunset. Figure 4.12 looks down on the North Pole. Recalling that the Earth rotates counterclockwise, you can deduce that it is sunset at Position C and sunrise at Position D. Figure 4.13 shows that any observer on the Earth who is looking in the direction of the new moon must also be looking in the direction of the Sun. Because the Sun and the new moon lie in the same direction in space, they rise and set together. Thus, the **new moon** rises at sunrise, sets at sunset, and is highest in the sky at noon.

The full moon lies in the opposite direction from the new moon and the Sun, as seen from Earth. When the Sun is rising on the eastern horizon, the full moon appears in the opposite direction (on the western horizon) where it is setting. Since the **full moon** is opposite the Sun, it rises at sunset and sets at sunrise.

The **first quarter moon** comes after the new moon but before the full moon. It rises after sunrise (when the new moon rises), but before sunset (when the full moon rises)—the first quarter moon rises at noon. Using similar reasoning, the first quarter moon should set around midnight while the **last quarter moon** should rise at midnight and set at noon.

FIGURE 4.13

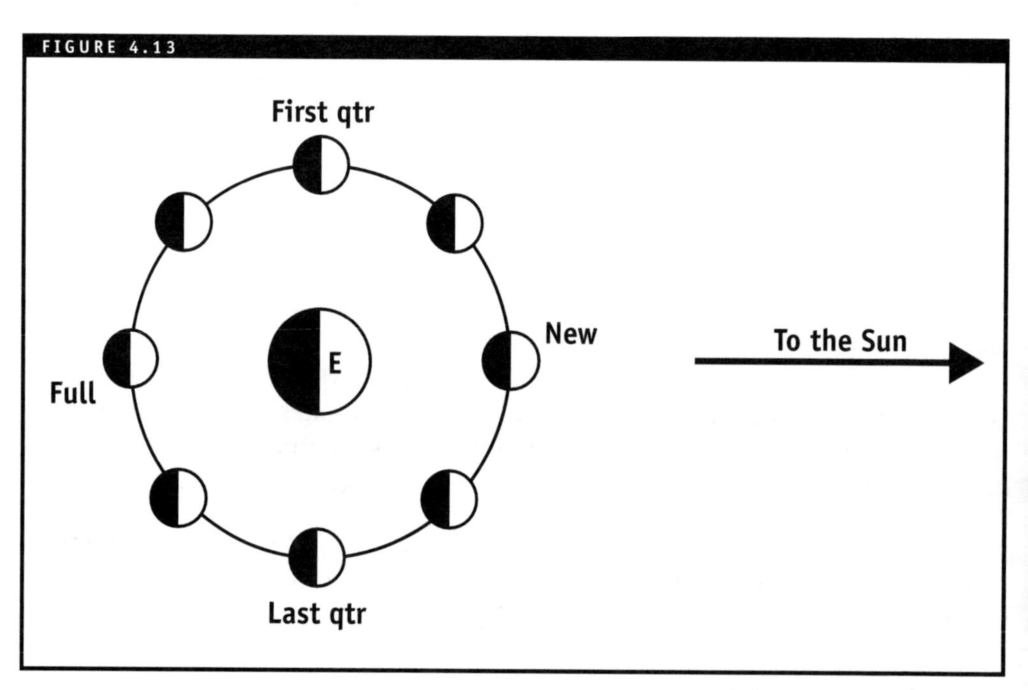

Viewing the Moon

FIGURE 4.14

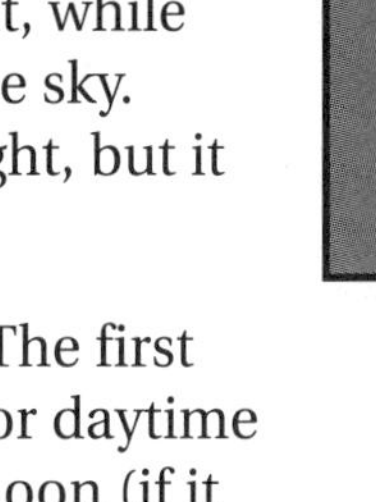

The Moon is above the horizon half the time, with the daytime and nighttime getting equal shares of the Moon's time. However, they do not get equal shares of the Moon's light. The Moon's constant motion and changing appearance make it an easy nighttime target and a sometimes challenging daytime object. The full moon is very bright and easy to see at night, while the dark new moon is nearly impossible to see in the bright daytime sky. Thus, the Moon is more obvious when it is above the horizon at night, but it can also be seen in the daytime sky if the phase is right.

The full moon will not be seen during the day, as it sets at sunrise. The first quarter moon will not be seen before noon. The best possibilities for daytime viewing of the Moon in the morning include the waning gibbous moon (if it is not so close to full as to have already set), the last quarter moon, and the waning crescent moon (unless it is too close to new to be visible).

In the morning, the **waning gibbous moon** is found in the western sky, away from the Sun, while the **waning crescent** is closer to the Sun but still west of it. Afternoon viewers should be looking for a **waxing gibbous moon** in the eastern sky or a **waxing crescent moon** closer to the Sun.

Earthshine

At certain phases, the dark portion of the Moon gives off a soft glow, allowing the full circle of the Moon to be seen (Figure 4.14). There may even be enough light to allow surface features to be distinguished in that region. The observed light is seen only in the crescent phases, in which a substantial portion of the darkened Moon is turned toward the Earth.

Sunlight shines on both the Moon and Earth, illuminating half of each. The Earth absorbs most of the light but reflects about a third of the incident rays back into space. Some of these reflected rays travel to the Moon and land on the dark side, where the Moon's extremely dark surface absorbs over 90% of them. Of the small fraction that is reflected from the Moon, some rays find their way back to Earth, as shown in Figure 4.15. If they fall on the daytime side of the Earth, they will not be noticed, due to the bright daytime sky. However, if these twice-reflected rays strike the dark side of the Earth, they may be seen against the darkened evening (or morning) sky. This phenomenon is known as **earthshine**; it can be observed on crescent moons in the morning or evening twilight.

FIGURE 4.15

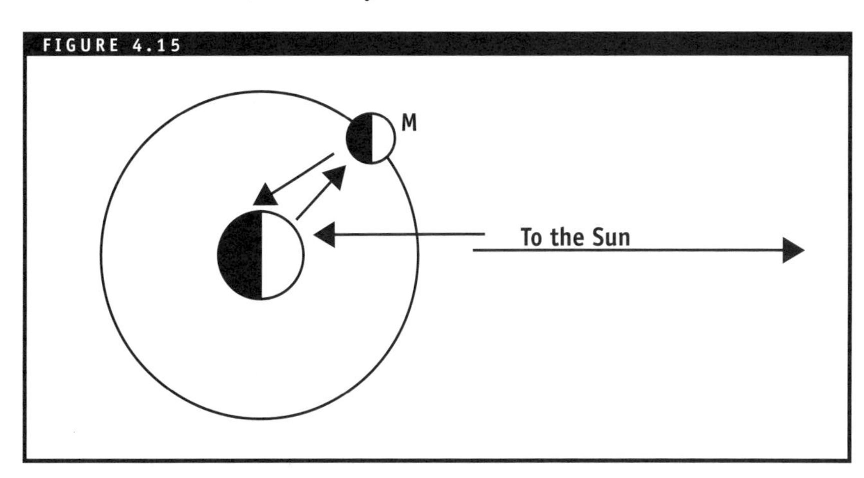

Chapter 5
Eclipses

Great Eclipse July 11, 1991

An **eclipse** occurs when one celestial body blocks light from reaching another. Eclipses are interesting phenomena for several reasons—they involve familiar bodies (the Earth, Moon, and Sun); they can be viewed easily by the public, without special equipment; they occur in a number of variations; and they are regular enough to be predictable, but rare enough to be special events.

FIGURE 5.1

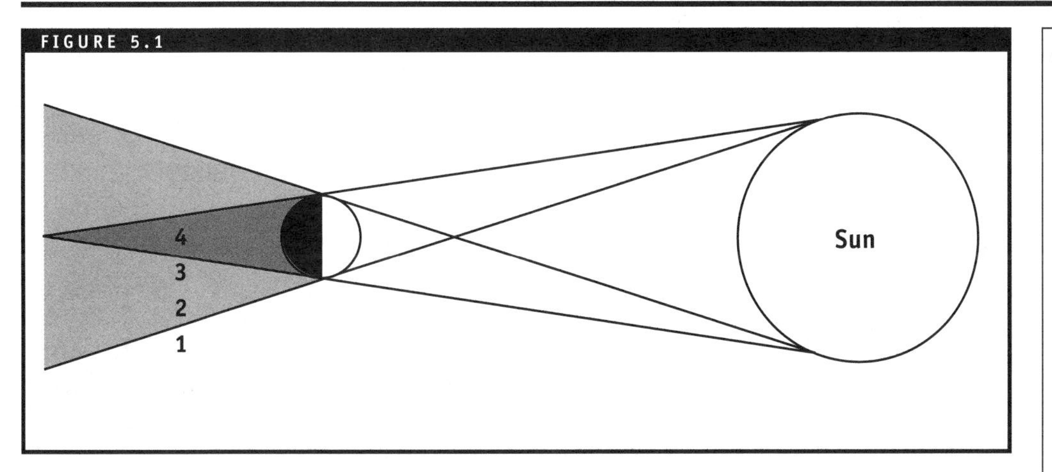

eclipse
: the darkening of a celestial body as it passes through the shadow of another body; the obscuration of all or part of the sun by a celestial body

lunar eclipse
: an eclipse in which parts of the Moon are darkened by the Earth's shadow; lunar eclipses occur at times of the full moon

penumbra
: the light shadow outside the dark shadow of an opaque body, where the light from the source of illumination is partially cut off

solar eclipse
: an eclipse in which parts of the Sun are hidden from view by the Moon; solar eclipses occur at times of the new moon

umbra
: the dark shadow of an opaque body, where direct light from the source of illumination is completely cut off

Shadows in Space

Normally, your shadow is visible only when it falls on the ground, a wall, or another object. Shadows in space are similar—you do not see them until they land on something. To make a shadow, you need two things—a source of light, and a body which is not a source of light. To see the shadow, you will also need a third body to intercept the shadow.

The obvious source of light in our Solar System is the Sun. The light that we see from the Earth, the Moon, and the planets is reflected sunlight. The Sun illuminates half the Earth and half the Moon at any given time. The night side of each body lies in the shadow produced by that body. But there is more to the shadow than just the night side.

To create Figure 5.1, draw rays from the top and bottom of the Sun through the top and bottom of a non-luminous body. Beyond this body, these rays form the boundaries of the shadow in space. Astronauts in a rocket traveling in space at Point 1 would be able to see the entire disk of the Sun, because rays drawn from Point 1 to any part of the Sun do not intersect the non-luminous body. At Point 2, however, the body blocks out rays from the top of the Sun, and the Sun would appear to have a small bite out of its upper edge. As the rocket moves from Point 2 to Point 3, the body blocks out even more of the Sun. This lightly shaded area that contains Points 2 and 3 is the part of the shadow we call the **penumbra**.

As the rocket moves on to Point 4, the astronauts find that the body blocks out all of the Sun's disk from view. This more darkly shaded area is called the **umbra**, and it is the darker of the two shadow regions. Eclipses occur when a third body (in this case, the rocket) moves into the penumbra or the umbra. Different types of eclipses occur depending on what body casts the shadow, what body intercepts the shadow, and what part of the shadow is intercepted.

We on Earth classify eclipses into two basic types—lunar and solar. In a **lunar eclipse**, parts of the Moon are darkened by the Earth's shadow, while in a **solar eclipse**, parts of the Sun are hidden from view by the Moon.

Lunar Eclipses

In a lunar eclipse, the Moon passes through the Earth's shadow. This means that the Moon must be on the same side of the Earth as the shadow, which is opposite the Sun. For this reason, a lunar eclipse requires a full moon. Because the Moon is much smaller than the Earth, it is able to fit entirely within the Earth's **umbra**, as shown in Figure 5.2 (the Sun is not shown). When this occurs, the Moon is completely immersed in the Earth's umbra, where practically no light can reach it. The Moon appears very dark compared to its normal full phase; such an eclipse is called a **total lunar eclipse**.

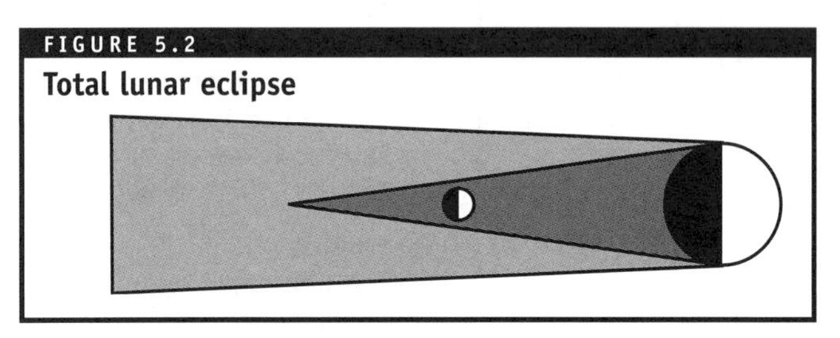

FIGURE 5.2
Total lunar eclipse

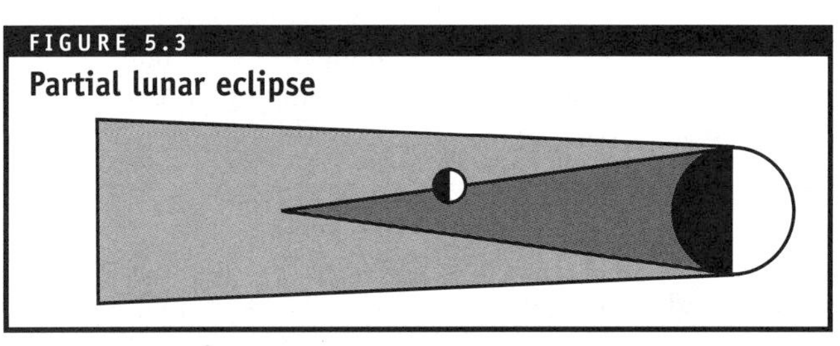

FIGURE 5.3
Partial lunar eclipse

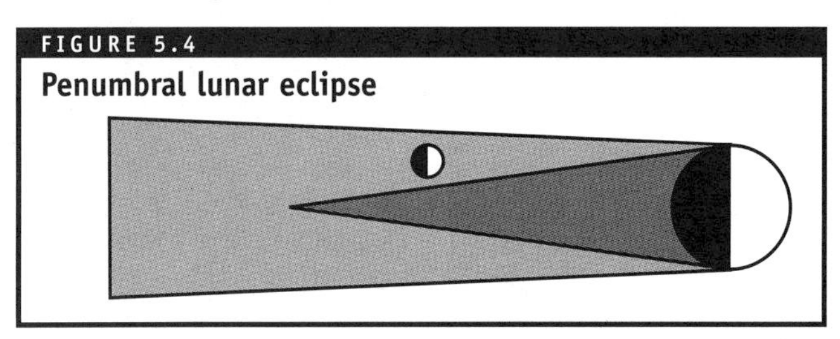

FIGURE 5.4
Penumbral lunar eclipse

The totally eclipsed Moon does not become black and invisible, despite being completely inside the Earth's umbra. It still receives light from the Sun which has been refracted, or bent, into the umbra by its passage through the Earth's atmosphere. Sunlight contains all the colors of the rainbow—red, orange, yellow, green, blue, indigo, violet—which normally all mix together to produce white light. The atmosphere tends to scatter, or deflect, the blue colors more efficiently than the red ones. The blue colors are easily removed from the Sun's rays to make blue sky here on the Earth; those rays that manage to pass through the atmosphere to be refracted into the umbra are predominantly red. For this reason, the totally eclipsed Moon is usually reddish in color and quite visible, though still much darker than a full moon.

Figure 5.2 is a side view of the Earth and Moon. The North Pole is toward the top of the Earth, and the Moon orbits the Earth in a plane **perpendicular** to the page. At the time of the eclipse, the Moon is coming out of the paper toward you; it will then move to the right between the Earth and you, and pass back through the paper to the right of the Earth.

As the Moon circles around in its orbit in this manner, it does not always pass directly through the umbra, as shown. Sometimes it only skims through the umbra, never becoming completely immersed; in this case, only part of the Moon is darkened by the umbra. Because the spherical Earth casts a circular shadow, the Moon will appear as a "bright cookie" with a bite out of it. Such an eclipse, shown in Figure 5.3, is called a **partial lunar eclipse**.

FIGURE 5.5

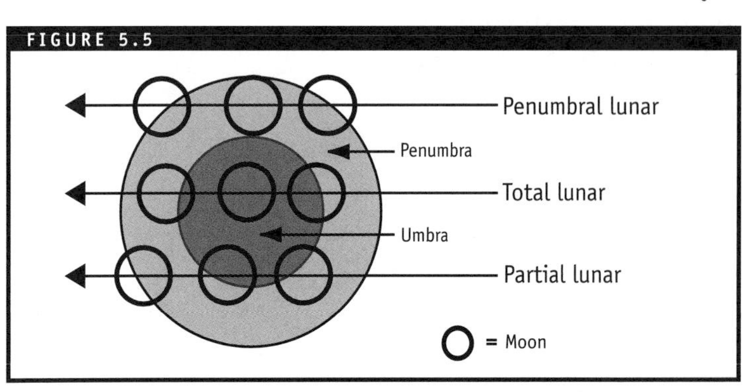

Sometimes the orbiting Moon will pass only through the penumbra, as shown in Figure 5.4. During the event, called a **penumbral lunar eclipse**, the Moon is not darkened appreciably. It appears to be shaded across its disk.

Figure 5.5 shows the changing appearance of the Moon as it moves through the Earth's shadow in each of the three lunar eclipses. Note that a total lunar eclipse has both partial and penumbral phases, while a partial eclipse also contains penumbral phases. The umbra and penumbra are shown here as they would appear if a huge projection screen were inserted at the distance of the Moon. The umbra and penumbra are usually not so apparent.

Lunar eclipses are fairly easy to see from Earth—as easy as the full moon. Any observer located on the dark side of the Earth (without too many clouds) will be able to view a lunar eclipse, which may last up to three hours, depending on the path that the Moon takes through the shadow.

Solar Eclipses

In a solar eclipse, the Moon blocks part of the Sun from our view, because it is between the Earth and the Sun, a condition that requires a new moon phase. Because the Moon is smaller than the Earth, its umbra is not big enough to contain the whole Earth. Instead, the Moon's umbra and penumbra will form darkened regions on the Earth's surface, as shown in Figure 5.6. The type of solar eclipse seen will depend on where the observer is located with respect to the shadow.

A person standing where the Moon's umbra touches the Earth (the black ellipse in Figure 5.6) would see the entire disk of the Sun obscured by the Moon. This spectacular event is called a **total solar eclipse**. The Moon covers the Sun's bright disk, leaving only its faint outer atmosphere (the **solar corona**) protruding from behind to form a "crown" around the new moon.

People standing in the adjacent shaded region where the Moon's penumbra touches the Earth would observe only part of the Sun's disk being covered by the Moon; they would see a **partial solar eclipse**—another bright cookie with a bite out of it. Those persons on the Earth's surface in the region outside the Moon's shadow would see no eclipse at all.

partial lunar eclipse
an eclipse in which only part of the Moon enters the umbra and is darkened

partial solar eclipse
an eclipse in which some of the Sun is blocked from view by the Moon

penumbral lunar eclipse
an eclipse in which the Moon passes through the penumbra and appears to be shaded across its disk, not significantly darkened

perpendicular
meeting at right (90°) angles

solar corona
a faint halo of gases surrounding the Sun visible during a total solar eclipse

total lunar eclipse
an eclipse in which the Moon is completely immersed in the Earth's umbra, where practically no light can reach it

total solar eclipse
an eclipse in which all of the Sun's disk is blocked from view by the Moon

umbra
the dark shadow of an opaque body, where direct light from the source of illumination is completely cut off

FIGURE 5.6

Total solar eclipse—Moon at perigee

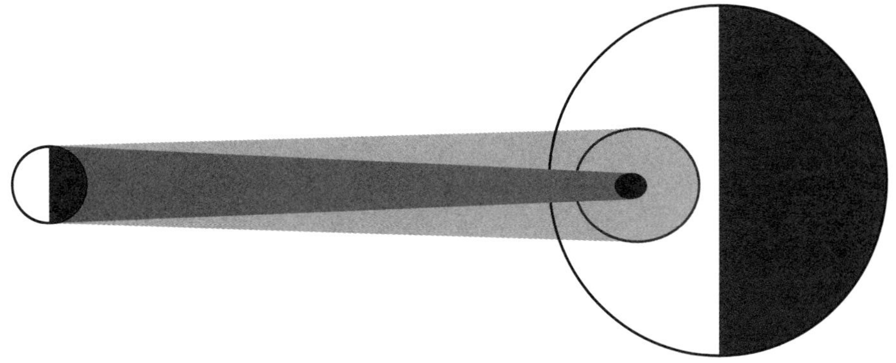

annular solar eclipse
a solar eclipse in which the Sun's disk appears as a ring around the new moon

annulus
ring shape

apogee
the point in an object's orbit around the Earth where it is farthest from the Earth

eclipse season
a month-long time interval when eclipses occur at new or full moons; normally, two eclipse seasons occur each year, about six months apart

eclipse track
the path of the Moon's shadow along the surface of the Earth during a solar eclipse; people living along the eclipse track will see a solar eclipse when the shadow passes over them

node
the points where the moon passes through the Earth's orbital plane

perigee
the point in an object's orbit around the Earth where it is closest to the Earth

FIGURE 5.7

Partial solar eclipse—Moon at apogee

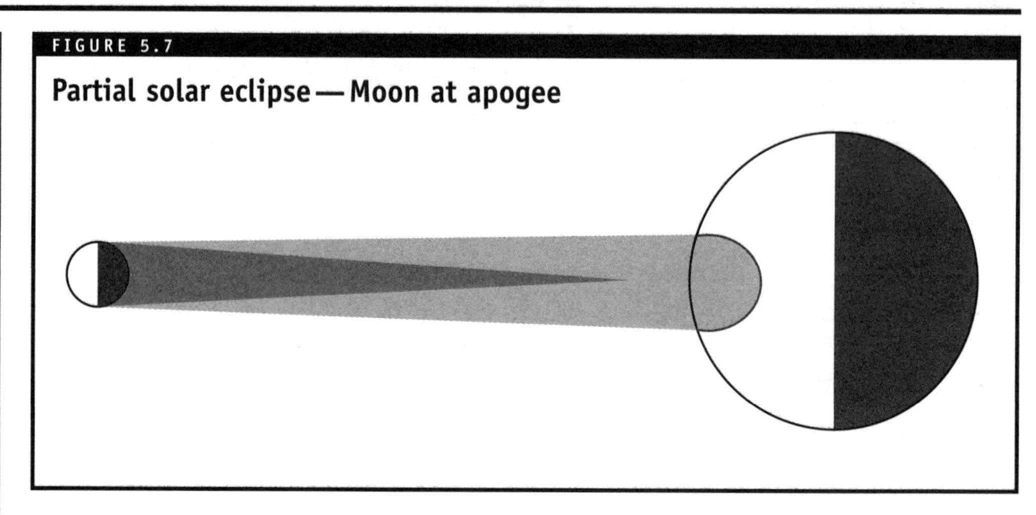

There is a third type of solar eclipse caused by the Moon's orbit being elliptical, not circular. Since the Earth's orbit is also elliptical, the Earth–Moon distance varies, from a minimum at **perigee** to a maximum at **apogee**. Figure 5.6 shows the conditions around perigee, which produce a total solar eclipse. Figure 5.7 shows the conditions at apogee, when the Moon is farther away from the Earth than normal. In fact, it is so far away that the Moon's umbra does not reach to the Earth's surface, making a total solar eclipse impossible.

A person standing in the shaded region where the Earth intercepts the Moon's penumbra would see a partial solar eclipse, as before.

However, a person standing directly below the tip of the Moon's umbra, exactly in line with the centers of the Moon and Sun, where the umbra would touch the Earth if it were long enough, would have a different view of the eclipse. The Moon and Sun would be aligned, but the Moon would be too small (due to its greater distance from Earth) to completely cover the Sun's disk. The Sun would still be visible as a bright ring of light around the dark Moon, as shown in Figure 5.8. Because a ring shape is called an **annulus**, this event is called an **annular solar eclipse**.

Just as the Moon moves through the Earth's shadow during a lunar eclipse, the Moon also moves in its orbit during a solar eclipse, dragging its own shadow across the surface of the Earth and producing an **eclipse track**. People living along the eclipse track will be treated to a solar eclipse when the shadow passes over them. Those who live outside the eclipse track will have to travel to it in order to view the eclipse in person. Because the Moon's umbra is not very wide where it touches the Earth's surface, total solar eclipses are very brief compared with total lunar eclipses. The maximum duration of totality is only a few minutes, but the partial phases may last for two or three hours.

FIGURE 5.8

Annular solar eclipse

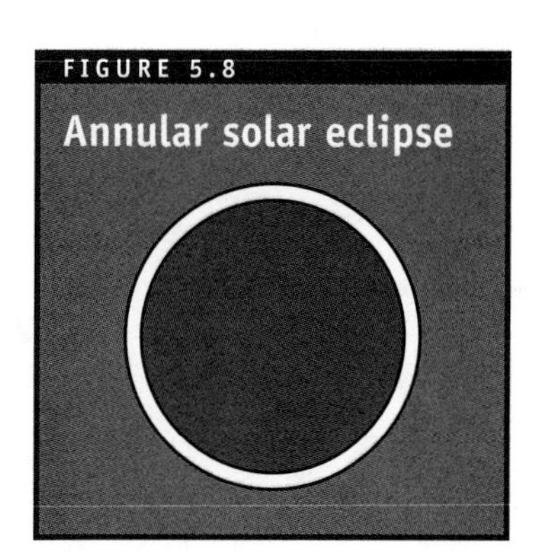

FIGURE 5.6

Total solar eclipse—Moon at perigee

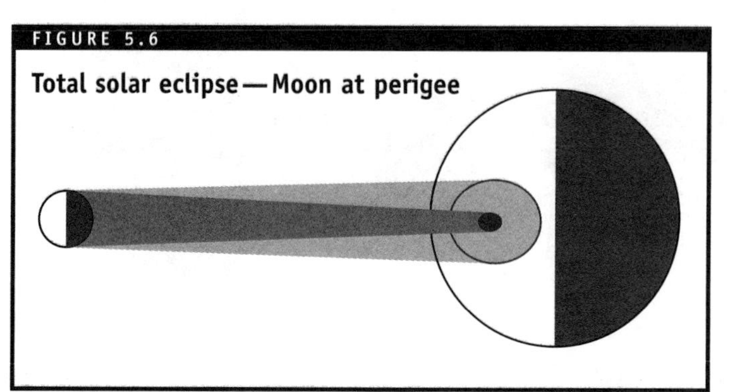

Eclipse Frequency

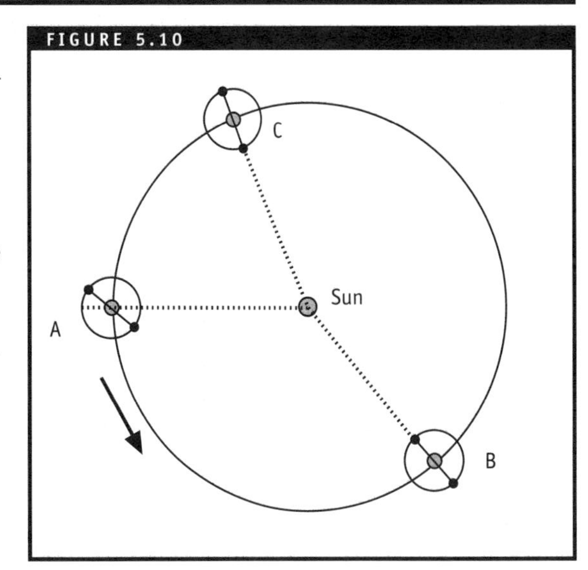

Eclipses require either a full moon (lunar eclipses) or a new moon (solar eclipses); additionally the Earth, Moon, and Sun must be almost perfectly in line. This is not common, as the Moon's orbit around the Earth is not in the same plane as the Earth's orbit around the Sun but is inclined to it by about 5°. When full, the Moon may pass up to 5° north or south of the Earth's shadow, missing it entirely. When new, the Moon may pass up to 5° north or south of the Sun, and its shadow won't usually fall on the Earth.

The alignment conditions necessary to create eclipses are illustrated in Figure 5.9, which shows a side view of the Earth-Moon system. Although the Moon on the left is full (as viewed from Earth), it is not quite in line with the Earth and Sun, and it misses the Earth's shadow. The Moon on the right is new (as viewed from Earth), but due to the inclination of the Moon's orbit (represented by the line joining the two phases), its shadow fails to fall on the Earth, and no eclipse occurs.

Eclipses occur at only two intervals during the year; these are known as **eclipse seasons**. At these times, the new and full moon will be close enough to the plane of the Earth's orbit that the Moon's shadow will fall on the Earth at new moon, and the Earth's shadow will fall on the Moon at full moon, causing solar and lunar eclipses, respectively. These eclipse seasons are not fixed in time, but migrate gradually around the calendar. Eclipses can occur in any month, but within a given year they will be confined to two month-long periods about six months apart.

Figure 5.10 shows the Earth's orbit around the Sun and the Moon's orbit around the Earth at three positions (A, B, and C). Because the Moon's orbit is inclined by 5° to the Earth's orbit, the two orbits are really not in the same plane. Half of the Moon's orbit is south of the Earth's orbital plane (behind the page), and half is north of it (in front of the page). The Moon passes through the Earth's orbital plane at the points marked by the small black dots. These points are called the **nodes** in the Moon's orbit. In order for an eclipse to occur, the Moon must be either new or full and must also be at or very near a node.

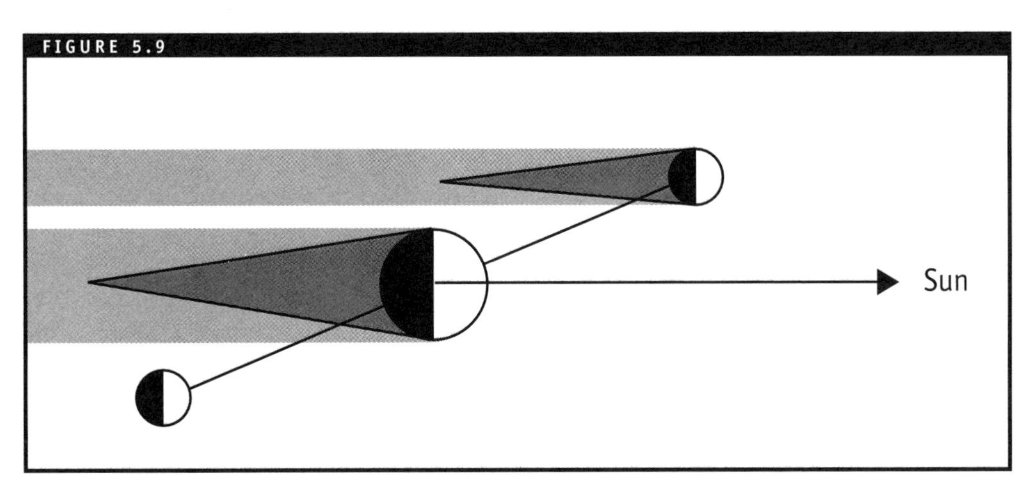

node
the points where the moon passes through the Earth's orbital plane

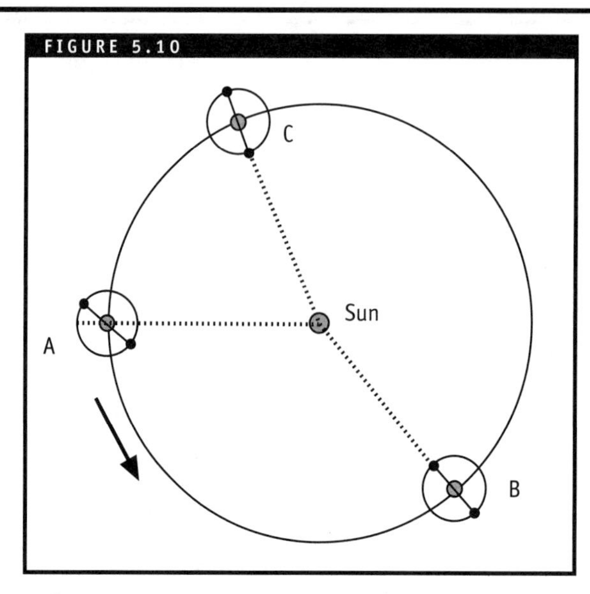

When the Earth is at Position A, neither the new moon nor the full moon would occur at a **node**. When the Moon does reach a node, it would be either a waxing gibbous or a waning crescent phase. Thus, no eclipses can occur at Position A.

At Position B, the Earth has moved in its orbit such that the line from the Sun to the Earth passes through the nodes. A new or full moon will occur at or near a node, resulting in eclipses. Position B marks the center of an eclipse season.

The next eclipse season will occur on the other side of the orbit, at Position C, which is not exactly 180° away from Position B. This is because the line joining the nodes is not fixed in direction; it slowly rotates clockwise, causing the interval between eclipse seasons to be somewhat less than six months.

Eclipse Information
The next two total solar eclipses to be seen from the continental United States will occur in 2017 and 2024. Those eager to view a total solar eclipse before that must be willing to travel out of the country. Find out more about eclipses and when they will happen on the World Wide Web.

http://sunearth.gsfc.nasa.gov/eclipse/eclipse.html

Viewing Solar Eclipses

Viewing a solar eclipse is somewhat hazardous because the Sun's rays can burn the unprotected eye. Telescopes or binoculars are particularly dangerous because they concentrate the Sun's rays like a magnifying glass. One technique is very simple: Punch a pinhole in a piece of cardboard and project the Sun's image onto a white screen or paper. *Never* look at the Sun directly.

The following page contains instructions on how to make a solar viewing tube, safe when used as instructed to view the Sun during an eclipse or to look at the Sun any time.

Solar Eclipse Viewing Tube

Materials needed:

Long cardboard tube

Aluminum foil

Pin (to make a pinhole)

White cardboard

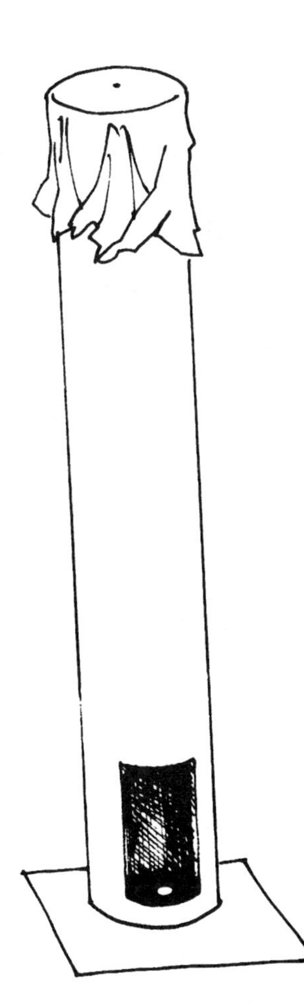

Cover one end of the cardboard tube with aluminum foil and make a pinhole in the foil. Cut a viewing port on the side of the tube near the opposite end. Cover that end (but not the port) with white cardboard.

To use a solar eclipse viewing tube:

1. Stand with your back to the Sun. *Do not* look at the Sun.
2. Hold the port end in one hand and rest the foil end on your shoulder.
3. Move your hand until the sunlight shines directly down the tube and the Sun's image appears on the white cardboard, as seen through the viewing port.

How will the shadow of the tube look when the image is visible through the port?

Explore the effects each of these has on the Sun's image:

❍ the diameter of the tube

❍ the length of the tube

❍ the size of the pinhole

Chapter 6

The Solar System

Voyager

PHOTO COURTESY OF JPL

The planets in our Solar System are Mercury, Venus, Earth, Mars, Jupiter, Saturn, Uranus, Neptune, and Pluto. Because all planets orbit the Sun in approximately the same plane, the Solar System is often said to be disk-shaped. The orbits of all the planets are elliptical—some more circular and others more elongated than Earth's. Most planet orbits do not overlap or intersect. (See the discussion on Pluto and Neptune on page 63, The Farthest Planet from the Sun.)

The planets move with respect to the Sun and the stars. In order to locate them, we must understand how their motion around the Sun affects how we view them from Earth. The beginning astronomer who understands planet motions can locate and observe planets in the sky with ease.

FIGURE 6.1

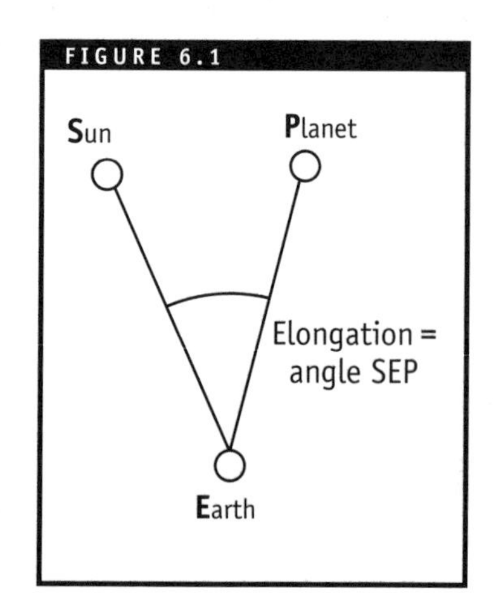

Planetary Configurations

Because the planets move around constantly, special terminology is used to note their positions with respect to the Sun and Earth. This section discusses the planetary **configurations**, which are frequently used in newspapers and almanacs to report the planets' positions.

Elongation

The key concept in describing planet positions is **elongation**, the angle between the Sun and the planet as seen from Earth. If the planet is *west* of the Sun, as shown in Figure 6.1, it has a *western* elongation; if the planet is *east* of the Sun, it has an *eastern* elongation. Elongations are generally given in degrees.

Inferior and Superior Planets

We divide the planets into two groups, depending on the locations of their orbits. Planets with orbits inside the Earth's are called **inferior planets**. These include Mercury and Venus. The remainder of the planets, with orbits lying outside the Earth's, are called **superior planets**. Figure 6.2 shows the configurations of the inferior planets.

Configurations of Inferior Planets

The term **conjunction** refers to two objects appearing close together in the sky, at or near 0° elongation. At **superior conjunction (SC)**, the planet is in line with the Sun on the opposite side of the Sun from us and, although fully illuminated, is not visible from Earth. At **inferior conjunction (IC)**, the planet is in line with the Sun between us and the Sun, with its illuminated side turned away from us, making this configuration even more difficult to observe. The planet can be best viewed from Earth when it is the most separated from the Sun—that is, when it has the **greatest elongation**, the maximum angle from the Sun. This occurs twice during each orbit, at **greatest eastern elongation (GEE)** and **greatest western elongation (GWE)**. At GEE, the planet is as far east of the Sun as it can get; it rises after the Sun, follows the Sun as it moves westward across the sky, and sets after the Sun sets. Thus, a planet at this configuration is best viewed just after sunset in the evening sky. At GWE, the planet is as far west of the Sun as it can get; it rises before the Sun, precedes the Sun westward across the sky, and sets before the Sun sets. Thus, a planet at this configuration is best viewed just before sunrise in the morning sky. At both greatest elongations the planet will be only half illuminated (quarter phase).

FIGURE 6.2

Inferior planets

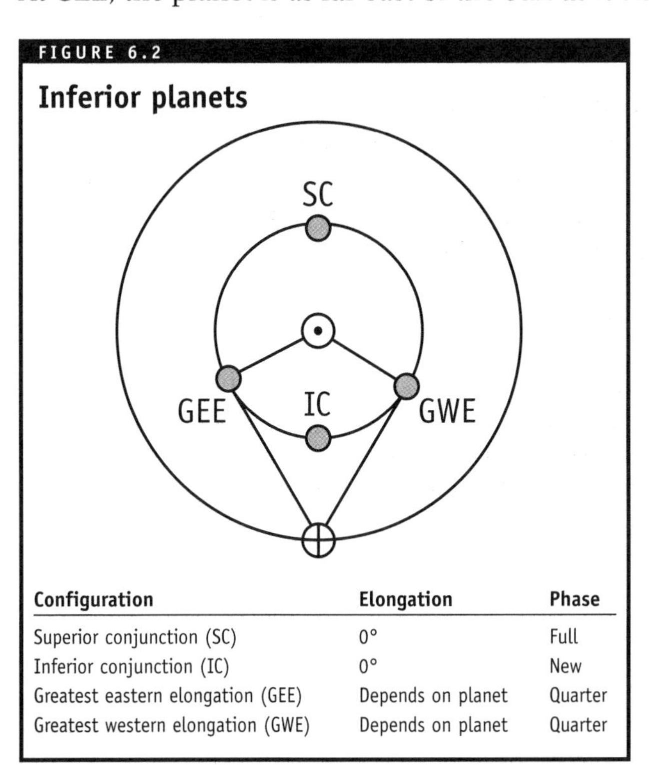

Configuration	Elongation	Phase
Superior conjunction (SC)	0°	Full
Inferior conjunction (IC)	0°	New
Greatest eastern elongation (GEE)	Depends on planet	Quarter
Greatest western elongation (GWE)	Depends on planet	Quarter

configuration
the relative position of bodies in space

conjunction
configuration in which two objects appear close together in the sky, at or near 0° elongation

elongation
the angle between the Sun and a planet, as seen from Earth, generally measured in degrees

greatest eastern elongation
configuration in which a planet is as far east of the Sun as it can get; it rises after the Sun, follows the Sun as it moves westward across the sky and sets after the Sun sets; this planet is best viewed just after sunset in the evening sky; the planet will be only half illuminated

greatest elongation
the maximum angle from the Sun

greatest western elongation
configuration in which a planet is as far west of the Sun as it can get; it rises before the Sun, precedes the Sun westward across the sky, and sets before the Sun sets; planets at this elongation are best viewed just before sunrise in the morning sky; the planet will be only half illuminated

inferior conjunction
configuration in which a planet is in line with the Sun between the Earth and the Sun

inferior planets
planets with orbits inside the Earth's—Mercury and Venus

superior conjunction
configuration in which a planet is in line with the Sun on the opposite side of the Sun from the Earth

superior planets
planets with orbits lying outside the Earth's—Mars, Jupiter, Saturn, Uranus, Neptune, and Pluto

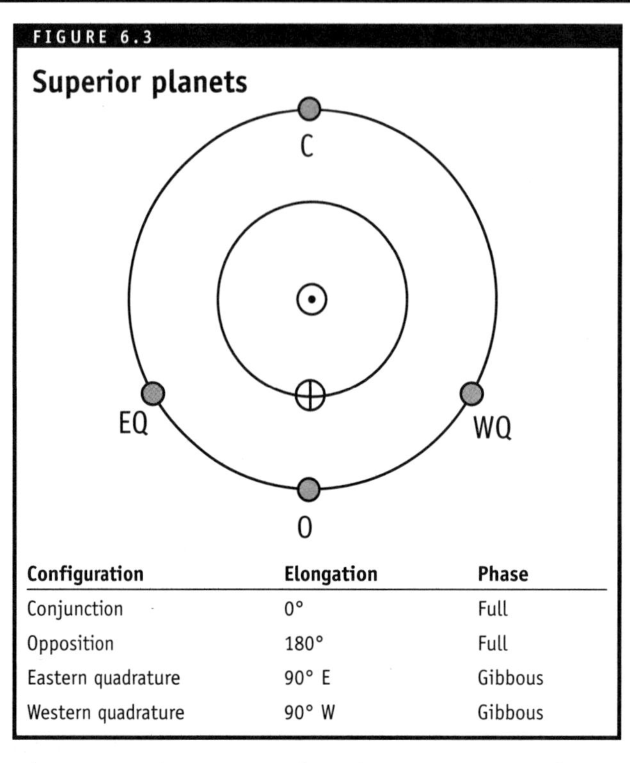

Configuration	Elongation	Phase
Conjunction	0°	Full
Opposition	180°	Full
Eastern quadrature	90° E	Gibbous
Western quadrature	90° W	Gibbous

We sometimes refer to morning planets and evening planets, meaning the time of day when the planet can be seen. A given planet is not always a morning planet or an evening planet, but switches back and forth as it orbits the Sun.

A handy way to remember which time is best for viewing the greatest elongations is to use the abbreviation itself. In GEE, the middle letter *E* stands for *eastern. Evening* and *eastern* both begin with the letter *e*. Use this to remind you that GEE planets are best viewed in the evening, and GWE planets are best viewed in the morning.

Configurations of Superior Planets

Figure 6.3 shows the configurations of the **superior planets**. With superior planets, there is only one way that conjunction can occur; a superior planet at conjunction is on the far side of the Sun from us and cannot be seen. **Opposition** refers to the planet's being on the opposite side of the Earth from the Sun. A planet in this position, like a full moon, will be fully illuminated and visible all night long.

Because opposition also marks the planet's closest approach to the Earth, this is the best configuration for viewing a superior planet. A planet at **eastern quadrature (EQ)** will be highest in the sky at sunset and visible in the evening sky, while a planet at **western quadrature (WQ)** will be highest at sunrise and visible in the morning sky. The two **quadratures** mark the phases of minimum illumination of the superior planet's surface, as seen from Earth. They occur at elongations of 90°.

To remember which are the best viewing times, apply the same mnemonic device used for the greatest elongations: use the abbreviation itself. The *E* in EQ stands for *eastern*. The *eastern* quadrature is best viewed in the *evening*—*eastern* and *evening* both begin with *e*. Planets at WQ, in the opposite direction, are best viewed in the opposite time of day, the morning.

astronomical unit (AU)
equal to about 93 million miles (150 million kilometers), the mean distance of the Earth from the Sun

eastern quadrature
configuration in which a planet has elongation = 90° E; will be highest in the sky at sunset and visible in the evening sky; will have a gibbous phase

eccentricity
the degree of flattening of an ellipse

opposition
configuration in which a planet is on the opposite side of the Earth from the Sun; a planet in this position behaves like a full moon, being fully illuminated and visible all night long; this is the best configuration for viewing a superior planet

quadrature
configuration of a superior planet in which its elongation is 90°; at eastern quadrature the planet is 90° east of the Sun, while at western quadrature the planet is 90° west of the Sun

superior planets
planets with orbits lying outside the Earth's—Mars, Jupiter, Saturn, Uranus, Neptune, and Pluto

synodic period
the time it takes for an inferior planet to lap the Earth or for the Earth to lap a superior planet

western quadrature
configuration in which a planet has elongation = 90°W; will be highest in the sky at sunrise and visible in the morning sky; will have a gibbous phase

Progress of the Planets

Once you understand the configurations, you can follow the progress of planets among the stars as they move from one configuration to the next. You will find it helpful to know how long a planet takes to move through its cycle of configurations.

The interval does not equal the planet's orbital period, because the Earth and the planet both move, and each configuration requires a particular alignment of the planet, Sun, and Earth.

Figure 6.4 shows the paths of the Earth and Mercury during the interval from one inferior conjunction to the next. Note that while the Earth moves about one third of the way around its orbit, Mercury moves one full orbit plus another third. The time it takes for an inferior planet to lap the Earth or for the Earth to lap a superior planet is called the **synodic period**. This is how long it takes for a planet to move through its cycle of configurations.

FIGURE 6.4

One synodic period:

inferior conjunction to inferior conjunction
or
opposition to opposition

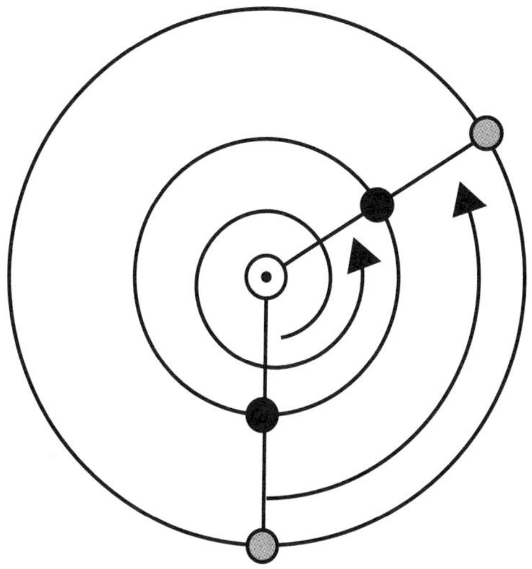

The Farthest Planet from the Sun

As mentioned in the beginning of this chapter, most of the planet orbits do not overlap or intersect, but Pluto and Neptune provide an interesting case study.

Pluto has the highest **eccentricity** (0.25) of any of the planets; its distance from the Sun varies from 49.2 AU at **aphelion** to 29.7 AU at **perihelion**.

Neptune has one of the lowest eccentricities (0.009) and a small variation, from 30.4 AU at aphelion to 29.8 AU at perihelion. Thus, Pluto at perihelion is closer to the Sun than Neptune!

aphelion
the point in an object's orbit around the Sun where it is farthest from the Sun

perihelion
the point in an object's orbit around the Sun where it is closest to the Sun

For 20 out of every 248 years in Pluto's period (most recently, from 1979 to 1999), Neptune is the most distant planet. Of course, since Pluto is more distant on the average, textbooks do not list Pluto as eighth and Neptune as ninth during this 20-year interval.

Despite the apparent crossing of their orbits, these two planets are not destined to collide. Pluto's orbit is inclined by 17° to the Earth's orbit, while Neptune's inclination is less than 2°. Because the orbits are in different planes, the paths of the two planets do not intersect. Nor do the two planets get close together. The periods of the two orbits are such that whenever Pluto reaches its perihelion, Neptune is always far away on the other side of its orbit.

FIGURE 6.5

Sidereal and synodic revolution periods of the planets (as seen from Earth)

Planet	Sidereal period	Synodic period
Mercury	88 days	116 days
Venus	225 days	584 days
Mars	687 days	780 days
Jupiter	11.86 years	399 days
Saturn	29.5 years	378 days
Uranus	84 years	370 days
Neptune	165 years	367.5 days
Pluto	248 years	366.7 days

The Closest Planet

Determining the closest planet can be a bit trickier. The planet closest to the Sun is always Mercury, because no planet approaches the Sun more closely, but the planet closest to Earth will be one of three planets (Mercury, Venus, or Mars), depending on how we measure the distance.

FIGURE 6.6

The closest planet

Planet	Distance from Sun (AU)	Minimum Distance from Earth (AU)	Maximum Distance from Earth (AU)
Mercury	0.38	0.62	1.38
Venus	0.72	0.28	1.72
Earth	1.0	0	0
Mars	1.5	0.5	2.5

Figure 6.6 shows the spacing of the orbits of the four inner planets, together with the minimum and maximum distances of each planet from the Earth. For simplicity, the orbits are considered to be circular.

A quick glance at the column headed *Minimum Distance from Earth* is enough to show that Venus' orbit lies nearest the Earth's. However, Venus gets far enough away that at times both Mercury and Mars can come closer to the Earth.

Figure 6.7 shows the orbits of the four inner planets around the Sun. In each diagram, the Earth is in the same position in its orbit, third from the Sun. Figure 6.7b shows that Venus can get closer to Earth than any other planet. However, Venus moves around in its orbit relative to Earth and is not always the closest planet to Earth, as shown in Figures 6.7a and 6.7c.

FIGURE 6.7

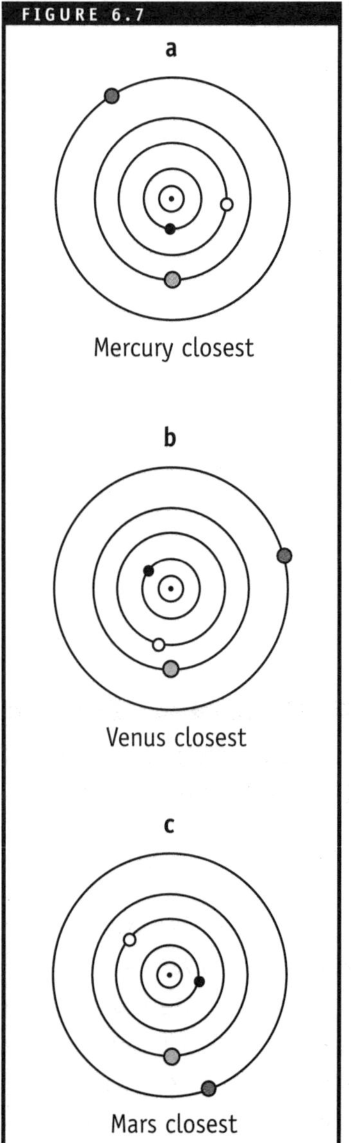

A Model Solar System

A simple way to present the most basic information about the Solar System, including names of the planets, their order from the Sun, and their relative sizes and distances from the Sun and from each other is in the form of a model.

To gain the most insight from the model, construct it to scale, such that each planet has the proper size relative to all the others. Choose any scale you want—the first model planet can be any size—but once the scale is chosen, the sizes of all the other model planets will be set.

Figure 6.8 provides information for one such model; it includes the Sun, the nine planets, and the Moon, along with their symbols and relative sizes. The scale has been chosen by selecting a blue marble as Earth, resulting in planet models that range in size from a 6-inch plastic foam ball for Jupiter, to a tiny ball for Pluto even smaller than the BB that represents the Moon. Using a collection of spheres is the best way to visualize the Solar System; but if you prefer, you can make a paper reproduction of the Solar System using the template on page 68. To create a reproduction of the Sun in the proper scale, use a rope 2½ feet long. By using one end as a centerpoint and the opposite end as a guide for a writing implement, draw a circle with a diameter of 5 feet.

Figure 6.8 also shows the distances from the model Sun and the orbital speed for each body.

The Solar System is often described as disk-shaped. However, the model Solar System, with a few tiny spheres scattered over several miles of emptiness, would hardly convey the image of a disk to the casual observer. The motions of the planets would slowly trace out the disk, but as we have seen, this motion is so gradual as to be almost imperceptible, especially for the outer planets.

FIGURE 6.8

Model Solar System

Body	Symbol	Ball	Diameter	Distance	Speed
Sun	☉	5-ft diameter ball	5 ft	n/a	n/a
Mercury	☿	Steel ball	5 mm	65 yards	15 feet/day
Venus	♀	White marble	½- in	125 yards	11 feet/day
Earth	⊕	Blue marble	½+ in	175 yards	9.6 feet/day
Earth's Moon	☽	BB	4 mm	16 inches (from Earth)	3.8 inches/day
Mars	♂	Red eraser	7 mm	265 yards	7.7 feet/day
Jupiter	♃	Plastic foam ball	6 in	½ mile	4.1 feet/day
Saturn	♄	Plastic foam ball	5 in	1 mile	3.1 feet/day
Uranus	♅	Plastic foam ball	2+ in	2 miles	2.2 feet/day
Neptune	♆	Plastic foam ball	2- in	3 miles	1.7 feet/day
Pluto	♇	Shiny ball	3 mm	4 miles	1.5 feet/day
Alpha Centauri		5-ft diameter ball	n/a	27,000 miles	n/a

Solar System Data

The Solar System in Figure 6.9 includes Earth diameters, distances from the Sun in astronomical units (AU), the time it would take light to travel that distance (light travel time), the orbital speeds, and the orbital periods. (Values for the Moon refer to its orbit around Earth.)

Figure 6.9 also includes one of the nearest stars to the Sun—Alpha Centauri. This star is a tremendous distance from the Sun—much farther away than Pluto, which appears quite local in comparison.

The more distant planets take longer to complete their orbits. This is because they travel farther (in larger orbits) and more slowly. The orbital period given for each planet is its sidereal revolution period, or sidereal year.

FIGURE 6.9

Solar System data

Body	Symbol	Earth Diameters	Distance (AU)	Light Travel Time	Actual Speed	Orbital Period
Sun	☉	109	n/a	n/a	n/a	n/a
Mercury	☿	0.38	0.38	3.2 minutes	48 km/sec	88 days
Venus	♀	0.95	0.72	6.0 minutes	35 km/sec	225 days
Earth	⊕	1	1	8.3 minutes	30 km/sec	365 days
Earth's Moon	☽	0.27	1/400	1.3 seconds	1.0 km/sec	27.3 days
Mars	♂	0.53	1.5	13 minutes	24 km/sec	687 days
Jupiter	♃	11.2	5.2	43 minutes	13 km/sec	11.9 years
Saturn	♄	9.4	9.5	1.3 hours	9.7 km/sec	29.5 years
Uranus	♅	4.0	19	2.6 hours	6.8 km/sec	84.0 years
Neptune	♆	3.8	30	4.1 hours	5.4 km/sec	165 years
Pluto	♇	0.18	39	5.5 hours	4.7 km/sec	248 years
Alpha Centauri		109	268,000	4.3 years	n/a	n/a

AU means Astronomical Unit

Bode's Law

Bode's Law was developed in the late 1700s as a way to represent the distances from the Sun of all the planets known at that time—Mercury through Saturn—and as a way to predict the positions of other bodies in the Solar System.

It is a simple formula that begins with the sequence of numbers *0, 1, 2, 4, 8, 16, 32 . . .* created by doubling each number after the 0 to get the next one. Each of these numbers is then multiplied by 3; next, add 4 to each product, and divide the sum by 10. The result gives the approximate distance of each planet in AU.

Step 1		Step 2		Step 3	Result	AU	Planet
0 x 3 =	0	+ 4 =	4	÷ 10 =	0.4	0.4	Mercury
1 x 3 =	3	+ 4 =	7	÷ 10 =	0.7	0.7	Venus
2 x 3 =	6	+ 4 =	10	÷ 10 =	1.0	1.0	Earth
4 x 3 =	12	+ 4 =	16	÷ 10 =	1.6	1.5	Mars
8 x 3 =	24	+ 4 =	28	÷ 10 =	2.8	–	–
16 x 3 =	48	+ 4 =	52	÷ 10 =	5.2	5.2	Jupiter
32 x 3 =	96	+ 4 =	100	÷ 10 =	10.0	9.5	Saturn

When Bode's Law was first published, there were only six planets known, but shortly after, in 1781, the planet Uranus was discovered. Amazingly, Uranus was found to fit Bode's Law.

Step 1		Step 2		Step 3	Result	AU	Planet
64 x 3 =	192	+ 4 =	196	÷ 10 =	19.6	19.2	Uranus

The law also predicted a planet at 2.8 AU, but none was known to be there. However, when the first asteroid, Ceres, was discovered in 1801, it was found to be at a distance of 2.8 AU, as were several others, which were discovered later.

Step 1		Step 2		Step 3	Result	AU	Planet
8 x 3 =	24	+ 4 =	28	÷ 10 =	2.8	2.8	Ceres (asteroid)

The discoveries of Neptune in 1846 and Pluto in 1930 showed that Bode's Law is not perfect—and not even very close for these outermost planets.

Step 1		Step 2		Step 3	Result	AU	Planet
128 x 3 =	384	+ 4 =	388	÷ 10 =	38.8	30.1	Neptune
256 x 3 =	768	+ 4 =	772	÷ 10 =	77.2	39.4	Pluto

What is the significance of Bode's Law? It appears to be just a curious mathematical relation that works pretty well to explain the location of most of the planetary orbits in our Solar System. It is not predicted by any scientific theory, and there is no reason to believe it will hold for other planetary systems.

Model Solar System Template

Trace planets on separate pages. Cut out. Arrange around Sun according to distances on figure 6.8, page 65.

Instructions for drawing the **Sun** (diameter 5 feet): Take a rope $2\frac{1}{2}$ feet long and use it as the radius. Hold one end of the rope at a center point and swing the rope around the center point to draw a circle with a diameter of 5 feet.

Jupiter

Saturn

Uranus

Neptune

Mars

Earth

Mercury

Earth's Moon

Venus

Pluto

Chapter 7

Planet Properties

Earthrise over the Moon PHOTO COURTESY OF NASA

The planets in our Solar System are convenient objects to study. We have sent unmanned probes to most of them and have obtained numerous pictures and considerable amounts of data. The goal of this chapter is to develop an understanding of the nature of each planet in the Solar System and the way in which it differs from the others. It also reviews a few other related bodies in the Solar System.

Planet Information
The Nine Planets
http://seds.lpl.arizona.edu/nineplanets/nineplanets/

Planets have a variety of properties. Some properties can be measured from Earth, while others require a close-up view. Some properties turn out to have different values later on. This means that in most listings of planet properties, there will be some numbers which are quite accurately known and some which are poorly known, with a full spectrum in between.

This book is no different—*just because you read it here or anywhere else does not mean that the number is absolutely correct.* Any two books on planets will likely disagree somewhat on the values for each planet, due to differing sources of information, changes in measurement techniques, different methods of rounding numbers, and of course, typographical errors. The values given here are intended to be reasonably correct and in general agreement with those found in most current sources.

Planetary Property Terms

Radius The *radius* is the distance from the center of the planet to the surface. If the planet is a perfect sphere, the radius will be the same all around. However, most of the planets are not perfect spheres; they have an equatorial bulge caused by their rotation. This means that the equatorial radius will be larger than the polar radius. Generally, when only one value for the radius is given, it is the equatorial radius. As explained in Chapter 1, from the radius we can also find the diameter ($2R$), circumference ($2\pi R$), surface area ($4\pi R^2$), and volume ($\frac{4}{3}\pi R^3$). Usually only the radius will be listed for each planet.

Oblateness The *oblateness* is used to measure the degree to which a planet is non-spherical. It is calculated by finding the difference between the equatorial and polar radii, and then dividing this difference by the equatorial radius. A perfect sphere has an oblateness of 0, while the most oblate planet in the Solar System has an oblateness of about 0.1.

Mass *Mass* is a measure of the amount of matter in a planet.

Density *Density* measures how tightly packed the matter is in a planet on average. It is the mass per unit volume, usually measured in grams per cubic centimeter. The density of water is one g/cc.

Composition Knowing the density of a planet gives us a clue to its *composition*—the types of matter that make up the planet's interior.

Surface gravity Gravity holds a planet together and holds objects onto the Earth's surface. The *surface gravity* depends on a planet's mass and radius and determines how much a person would weigh on the planet.

Escape velocity The *escape velocity* is the minimum speed needed to launch a rocket (or anything else) from the surface of a planet. Knowing the escape velocity allows you to determine how easily atmospheric particles can escape from the planet, and thus, what sort of atmosphere the planet retains.

Atmosphere A planet may have a layer of gases surrounding its surface. This *atmosphere* may contain clouds that prevent our viewing the surface, or it may be transparent. The gases that compose the atmosphere may be beneficial to life on the surface or poisonous.

Surface features If the atmosphere is transparent or nonexistent, a variety of *surface features* on a planet may be identified: craters, mountains, valleys, volcanoes, and so on. In some cases astronomers can identify these features despite thick cloud cover by using radar to probe beneath the clouds.

Sidereal rotation period The planets all rotate (spin) on their axes. The *sidereal rotation period* measures the duration of one rotation with respect to the stars—a relatively stationary reference frame. You can determine this period by going outside at night, pointing at a star, and measuring how long it takes for the same star to move around the sky and return to your finger. Most planets spin counterclockwise about their north poles. Planets that spin clockwise about their north poles are said to have *retrograde rotation.*

Sidereal revolution period Planets orbit around the Sun. The time for a planet to complete one orbit with respect to a distant star is the *sidereal revolution period.*

Solar day You can also measure the rotation of a planet with respect to the Sun. Go outside during the day, point at the Sun, and wait for one solar day until the Sun again lines up with your finger. In general, the *solar day* and the sidereal rotation period will not be identical, although they may be very close.

Obliquity A planet's rotation defines one axis, and its revolution defines another. The angle between these two axes is the *obliquity,* sometimes called the tilt. This angle is responsible for many of the Earth's seasonal effects.

Average distance from the Sun Planets move in elliptical orbits, constantly changing their distances from the Sun. The *average distance from the Sun* is a measure of the size of the planet's orbit.

Orbital eccentricity The shape of an elliptical orbit is measured by the *orbital eccentricity*, which ranges from 0 for a circular orbit to near 1 for a highly elliptical orbit.

Orbital inclination The Solar System is nearly planar, but not perfectly so. The orbits of the planets are all inclined to each other by small angles. The *orbital inclination* measures the angle between the planet's orbit and the ecliptic (the Earth's orbital plane).

Number of moons Most of the planets have *moons*, or natural satellites. The larger moons of each planet have been known for some time. Smaller ones are still being discovered, primarily by our planetary probes.

Rings Several of the planets have *rings*—systems of many tiny particles orbiting the planet to form a planar ring. Some of these rings can be seen easily from Earth, while others have only recently been discovered by spacecraft.

Planetary Groups

Some planets share similar properties. This allows us to group them together and make some general statements about each group. The two main groups are the *terrestrial planets* and the *Jovian planets*. The terrestrial—or Earth-like—planets are Mercury, Venus, Earth, and Mars; the Jovian—or Jupiter-like—planets are Jupiter, Saturn, Uranus, and Neptune. Pluto is not included in either group.

The table in Figure 7.1 shows the differences in several basic properties for the two groups. In each case the division is absolute—every terrestrial planet is more dense than any Jovian; every Jovian planet has more moons than any terrestrial; and so on. Pluto, on the other hand, shares some properties with each group, and thus is classed by itself.

FIGURE 7.1

Planetary groups

Property	Terrestrial planets	Jovian planets	
	Mercury, Venus, Earth, Mars	Jupiter, Saturn, Uranus, Neptune	Pluto
Distance from Sun	Close	Far	Far
Size/Mass	Small	Large	Small
Density	High	Low	Low
Composition	Rocks, iron	Gases, H, He	Ices
Revolution	Fast	Slow	Slow
Rotation	Slow	Fast	Slow
Moons	Few (3)	Lots (60+)	(1)
Rings	None	All	None

The source for information about the planets included in this section is Jeffrey K. Wagner's *Introduction to the Solar System* (Holt, Rinehart & Wilson, 1991). In many cases, values have been rounded to provide more convenient numbers. A table of properties and a brief discussion of each planet is given. As noted on page 69, not all sources will agree, and some properties may be revised by new observations.

Earth

Earth's Properties

Equatorial Radius
6,378 km (= 1 R_{earth})

Oblateness 0.00335

Mass 5.97 x 10^{27} grams
(= 1 M_{earth})

Density 5.52 grams/cc

Surface Gravity 980 cm/s^2
(= 1 g)

Escape Velocity 11.2 km/s
(≈ 25,000 mi/hr)

Sidereal Rotation Period
23.934472 hours

Solar Day 24 hours

Obliquity 23.45°

Sidereal Revolution Period
365.256 days

Average Distance from the Sun
149.6 million km (= 1 AU)

Orbital Eccentricity 0.017

Orbital Inclination (0°)

Number of Moons 1

Earth is listed first because we use it to define or derive several units, including those for time (years, days, hours, seconds), mass (M_{earth}), length (AU, R_{earth}), and gravity (g). Earth is the largest of the terrestrial planets and has the highest density.

It is the only planet with substantial amounts of liquid water on its surface and the only planet known to have life. This life has had a part in establishing the present atmosphere—the large fraction (nearly 21%) of highly reactive free oxygen in our atmosphere was produced by plant life over the last few billion years. No other planet in our Solar System has such a high abundance of O_2.

Earth is one of only two terrestrial planets to have a moon. The Moon's properties are listed here for comparison with those of the planets.

Full Earth, Africa PHOTO COURTESY OF NASA

Moon

The Moon is included here even though it is not a planet. It is close to us and easily studied using binoculars or a small telescope. One of its distinctive features is a lack of any significant atmosphere, so there are no clouds to obscure our view of the surface, where craters, mountains, and **maria**—huge lava-filled basins—are visible. Most of these features were created by the impacts of other celestial bodies on the lunar surface. Impacts of the larger bodies excavated the basins, which later were flooded with molten rock from the interior. Many of the mountain ranges seen today on the Moon are the rims of these huge basins. Numerous smaller impacts have formed the craters we find covering most of the Moon's surface. Rocks on the Moon are very old, approaching the 4.6 billion year age of the Solar System.

Craters are seen on bodies throughout the Solar System, indicating that cratering is a widespread phenomenon. Only a few craters are found on Earth, not because Earth has avoided collisions, but rather because the erosion and tectonic motions of Earth's crust have largely erased all but the most recent craters.

The lack of a lunar atmosphere has other consequences. It means that space suits are necessary for lunar visitors, for there is no air to breathe there. Lack of air also means there is no wind on the Moon, no scattering of light by air molecules (making skies black rather than blue), and no sound waves (making radio communication necessary). The Moon does have gravity, but it is only one sixth as strong as Earth's; a rock dropped on the Moon will still fall to the surface, but it will not fall as rapidly. With no air and no water, the Moon is a lifeless body.

Night and day on the Moon are fairly long—approximately two weeks each—as the rotation period with respect to the Sun is the synodic revolution period found for the Moon's phases (see pages 46 to 48). The Sun moves across the lunar sky, but the Earth does not. An astronaut on the near side of the Moon would always find the Earth above the horizon, during both day and night, and would occasionally be able to view solar eclipses as the Earth covers the Sun. An astronaut on the far side of the Moon would be unable to see the Earth or send radio messages to it.

Far side of the Moon PHOTO COURTESY OF NASA

The Moon's Properties

Equatorial Radius 1,738 km (= 0.272 R_{earth})

Oblateness 0.00063

Mass 7.35 x 10^{25} grams (= 0.0123 M_{earth})

Density 3.3 grams/cc

Surface Gravity 1/6 g

Escape Velocity 2.4 km/s

Sidereal Rotation Period 27.32 days

Solar Day 29.53 days

Obliquity 6.68°

Sidereal Revolution Period 27.32 days

Average Distance from Earth 384,400 km (= 1/389 AU)

Orbital Eccentricity 0.055

Orbital Inclination 5.15°

Number of Moons 0

maria
huge lava-filled basins on planetary surfaces

Mercury

Mercury's Properties

Equatorial Radius 2,439 km (= 0.382 R_{earth})

Oblateness 0

Mass 3.30 x 10^{26} grams (= 0.055 M_{earth})

Density 5.43 grams/cc

Surface Gravity 0.37 g

Escape Velocity 4.2 km/s

Sidereal Rotation Period 58.65 days

Solar Day 176 days

Obliquity 0.0

Sidereal Revolution Period 87.97 days

Average Distance from the Sun 0.387 AU

Orbital Eccentricity 0.206

Orbital Inclination 7.0°

Number of Moons 0

greenhouse effect
process by which a planetary atmosphere absorbs outgoing radiation, thus maintaining higher temperatures on the surface of the planet

maria
huge lava-filled basins on planetary surfaces

Mercury, the smallest terrestrial planet, is Moon-like in appearance, as shown in images from *Mariner 10* flybys in 1974 and 1975. It has no atmosphere and lots of craters, but relatively few of the **maria** found on the Moon. Its proximity to the Sun makes it difficult to study from the Earth, even though it is visible to the unaided eye.

Mercury has an interesting relation between its rotation and its revolution—the two motions are linked in a 3:2 resonance, which means that Mercury rotates three times in every two revolutions around the Sun. The solar-day length depends on both of these motions, resulting in an interval from one noon to the next on Mercury being equal to two revolutions, or 176 days. Thus, Mercury residents would experience two Mercury years per Mercury day. But with long, hot days and equally long, cold nights, Mercury is not apt to attract many visitors in the near future.

Mercury PHOTO COURTESY OF NASA

Venus

Venus is very similar in size to the Earth and has often been called Earth's twin, or sister, planet. Early observers knew that Venus was closer to the Sun—and therefore hotter—and covered with a layer of dense clouds. Assuming that the clouds were similar to those found on Earth, these observers envisioned a tropical climate for Venus, with good possibilities for a variety of life. Discovery of carbon dioxide in the atmosphere made the possibility of plant life on Venus even more likely.

We now know that the temperatures on Venus are far too high for liquid water to exist, due to the **greenhouse effect**. Light from the Sun filters through the clouds and warms the surface of the planet, which then reradiates the energy as infrared rays. However, the dense carbon dioxide atmosphere absorbs the infrared rays, trapping the energy below the cloud layers and keeping the surface temperature abnormally high—nearly 900° F.

In a greenhouse, light from the Sun filters through the glass and warms the interior, which then reradiates the energy as infrared rays. However, the glass does not allow infrared rays to pass through, thus trapping the energy inside the greenhouse and keeping the plants warm.

With its high temperatures and dense atmosphere, the surface of Venus is quite unlike that of the Earth, and certainly is unsuitable for life as we know it. Venus appears to be geologically active. Radar mapping of the cloud-covered surface by the *Magellan* spacecraft in 1991 revealed a variety of terrain, including impact craters and features produced by volcanic activity. The clouds are also unique. Unlike our clouds of water droplets, those on Venus are made of sulfuric acid droplets.

Like Mercury, Venus has a slow rotation; its 243-day sidereal rotation period is the longest of any planet. In addition, its rotation is retrograde, or backward, from the normal counterclockwise motion of the planets. This motion couples with the planet's orbital motion to produce a solar day on Venus of 117 days. Thus, even though Venus has the longest sidereal day, its solar day is second to Mercury's in length. And while Mercury calendars have two years per day, those on Venus have approximately two days per year.

Venus' Properties

Equatorial Radius 6,052 km (= 0.949 R_{earth})

Oblateness 0

Mass 4.87 x 10^{27} grams (= 0.815 M_{earth})

Density 5.24 grams/cc

Surface Gravity 0.876 g

Escape Velocity 10.4 km/s

Sidereal Rotation Period 243 days (retrograde)

Solar Day 117 days (retrograde)

Obliquity 2.7°

Sidereal Revolution Period 224.7 days

Average Distance from the Sun 0.723 AU

Orbital Eccentricity 0.007

Orbital Inclination 3.39°

Number of Moons 0

Venus

PHOTO COURTESY OF NASA

Mars

Mars' Properties

Equatorial Radius 3,393 km (= 0.532 R_{earth})

Oblateness 0.052

Mass 6.42 x 10^{26} grams (= 0.107 M_{earth})

Density 3.94 grams/cc

Surface Gravity 0.381 g

Escape Velocity 5.0

Sidereal Rotation Period 24.623 hours

Solar Day 24.660 hours

Obliquity 25.2°

Sidereal Revolution Period 686.99 days

Average Distance from the Sun 1.524 AU

Orbital Eccentricity 0.093

Orbital Inclination 1.85°

Number of Moons 2

The "red planet," Mars, has been studied with great interest for many years, due in part to its similarities to the Earth. Mars has a mostly transparent atmosphere, which allowed early astronomers to view seasonal changes in the polar ice caps and the blue-green markings on its otherwise orange surface. Such observations suggested the presence of water and perhaps plant life on Mars. Later, observations of straight-line markings and their interpretation as canals caused some to assume the presence of Martians, who supposedly constructed the canals to bring water from the polar regions to irrigate the equatorial deserts.

Numerous books and movies supported this idea, but the *Mariner* and *Viking* missions to Mars in the 1960s and 1970s did not. They depicted a less hospitable world, with a very thin atmosphere of carbon dioxide and temperatures too low for liquid water to exist. The Martian surface is certainly interesting, with occasional craters, inactive volcanoes, dust storms, dry riverbeds, and polar ice caps of water ice and frozen carbon dioxide (dry ice), but there is no indication of life on Mars. Although it appears that in the distant past the atmosphere was denser and liquid water probably flowed on the Martian surface, the *Viking* spacecraft that landed there in 1976 found no conclusive evidence of any type of life.

Mars has two of the three satellites that orbit terrestrial planets—Phobos and Deimos are tiny moons with irregular shapes, which may be captured asteroids. Because of Phobos' very rapid orbital motion around Mars, Martian observers would see Phobos rise in the west and set in the east, while Deimos follows a more normal east-to-west route.

Mars PHOTO COURTESY OF NASA

Jupiter

Jupiter is completely different from the terrestrial planets previously described; in fact, it gives the Jovian planets their designation. (*Jove* was another name for the Roman god Jupiter.)

Jupiter is the largest planet in the Solar System and also the most massive, containing more mass than all the other planets combined. It is a gas giant, composed largely of gases of hydrogen, helium, and other lightweight elements, with a density about one quarter the value of the Earth's. The planet probably has a small core of iron and rock at its center, which is surrounded by hydrogen and helium compressed to a liquid form by the tremendous pressures of Jupiter's interior.

The surface of Jupiter is not solid but fluid, making it quite unlike the terrestrial planets. We cannot see this surface because it is covered by layers of clouds; the patterns we see on Jupiter—colored bands, belts, and oval spots—are all features in the cloud layers. Jupiter's rapid rotation causes considerable motion in these features, with rotating spots, and adjacent bands flowing in opposite directions. The most notable of these features is the Great Red Spot, which has been observed for over three centuries.

Jupiter has a number of satellites, with the four largest (Io, Europa, Ganymede, and Callisto) easily visible with binoculars. Their orbital periods range from two days to about two weeks, and they can be seen to change position from night to night. Io and Europa are about the size of our Moon, while Ganymede and Callisto are about the size of Mercury. The *Voyager* flyby missions in 1979 discovered several new moons and a set of thin rings orbiting the planet.

Jupiter PHOTO COURTESY OF NASA

Jupiter's Properties

Equatorial Radius 71,400 km (= 11.19 R_{earth})

Oblateness 0.065

Mass 1.90 x 10^{30} grams (= 318 M_{earth})

Density 1.33 grams/cc

Surface Gravity 2.64 g

Escape Velocity 59.6 km/s

Sidereal Rotation Period 9.92 hours

Solar Day 9.92 hours

Obliquity 3.1°

Sidereal Revolution Period 11.9 years

Average Distance from the Sun 5.203 AU

Orbital Eccentricity 0.049

Orbital Inclination 1.3°

Number of Moons 16

Saturn

Saturn's Properties

Equatorial Radius 60,000 km (= 9.41 R_{earth})

Oblateness 0.108

Mass 5.69 x 10^{29} grams (= 95.2 M_{earth})

Density 0.70 grams/cc

Surface Gravity 1.15 g

Escape Velocity 35.6 km/s

Sidereal Rotation Period 10.5 hours

Solar Day 10.5 hours

Obliquity 26.7°

Sidereal Revolution Period 29.5 years

Average Distance from the Sun 9.555 AU

Orbital Eccentricity 0.056

Orbital Inclination 2.5°

Number of Moons 18

Saturn is the most distant planet known to ancient astronomers; its brightness makes it easily visible to the naked eye. In a small telescope, Saturn is a special sight because of its system of bright rings. These rings are made of millions of small particles orbiting the planet in its equatorial plane. Unlike the rings of the other Jovian planets, which are thin and relatively distinct, Saturn's rings appear broad and fairly continuous, without significant gaps between them.

In other respects, Saturn is another gas giant, smaller than Jupiter. It, too, is composed chiefly of hydrogen and helium, but due to its smaller mass, it is less compressed than Jupiter. As a result, Saturn has the lowest density of any planet—so low that Saturn could float in water if a large enough bathtub could be found. Saturn's clouds are not as colorful as Jupiter's; the bands and spots found here are much subtler. Saturn's rapid rotation has made it the most oblate planet, with a noticeable difference between its polar and equatorial diameters.

Saturn has many moons, as do all the Jovian planets. The majority of these appear to be solid, with surfaces heavily covered with craters. The largest moon—Titan—has its own atmosphere, which has prevented direct viewing of its surface and promoted speculation on the possibility of life having developed there. Most of the smaller moons were discovered by the *Voyager* spacecraft during flyby missions in 1980 and 1981.

Saturn

PHOTO COURTESY OF NASA

Uranus

Uranus was not studied by early astronomers because it is barely bright enough to be seen by the unaided eye and thus easily mistaken for a faint star. It was discovered in 1781 by William Herschel, who noted that its telescopic image appeared somewhat different from that of stars in the same field. Further observations showed that it was moving slowly among the stars, orbiting the Sun.

Uranus is one of the smaller Jovian planets, but it still has a diameter four times that of the Earth. Uranus is covered with a thick cloud cover and lacks a solid surface; its clouds have proved to be especially featureless, without significant spots or bands.

The most distinctive feature of Uranus has always been its obliquity of 82.1°, which means that the rotational axis lies nearly in the orbital plane. (To see this effect, tip a globe until its axis is almost horizontal. Such an orientation for Earth would result in dramatic seasonal changes.) At the solstices, Uranus' poles point almost directly toward the Sun, rather than just tilting toward it as do Earth's. The "land of the midnight sun" (see Chapter 3) extends from the pole to within 8° of the equator on Uranus, and the Sun reaches the zenith as far north as 8° from the pole! (Of course, the cloud cover would prevent any residents there from actually viewing these phenomena.)

Uranus has an 82° obliquity and retrograde (CW) rotation; some sources will list a 98° obliquity and normal (CCW) rotation. Both are valid—they result from using the two different poles to measure the angle and view the rotation.

Uranus' Properties

Equatorial Radius 25,400 km (= 3.98 R_{earth})

Oblateness 0.030

Mass 8.66 x 10^{28} grams (= 14.5 M_{earth})

Density 1.30 grams/cc

Surface Gravity 1.06 g

Escape Velocity 21.3 km/s

Sidereal Rotation Period 17.24 hours (retrograde)

Solar Day 17.24 hours (retrograde)

Obliquity 82.1°

Sidereal Revolution Period 84.0 years

Average Distance from the Sun 19.22 AU

Orbital Eccentricity 0.046

Orbital Inclination 0.8°

Number of Moons 21

In 1977, Uranus became the second planet known to have rings, but they are not easily seen from Earth. The passage of *Voyager 2* in 1986 increased the number of known moons of Uranus from five to 15; further analysis of the *Voyager* images and recent observations from Earth have added several more moons to the total. Both the rings and the moons orbit in Uranus' equatorial plane, which is tilted by 82° from its orbital plane. Sometimes when viewed from the Earth at certain points in its orbit, Uranus resembles a huge target, with the planet as the bull's-eye and the rings and moon orbits forming the surrounding circles.

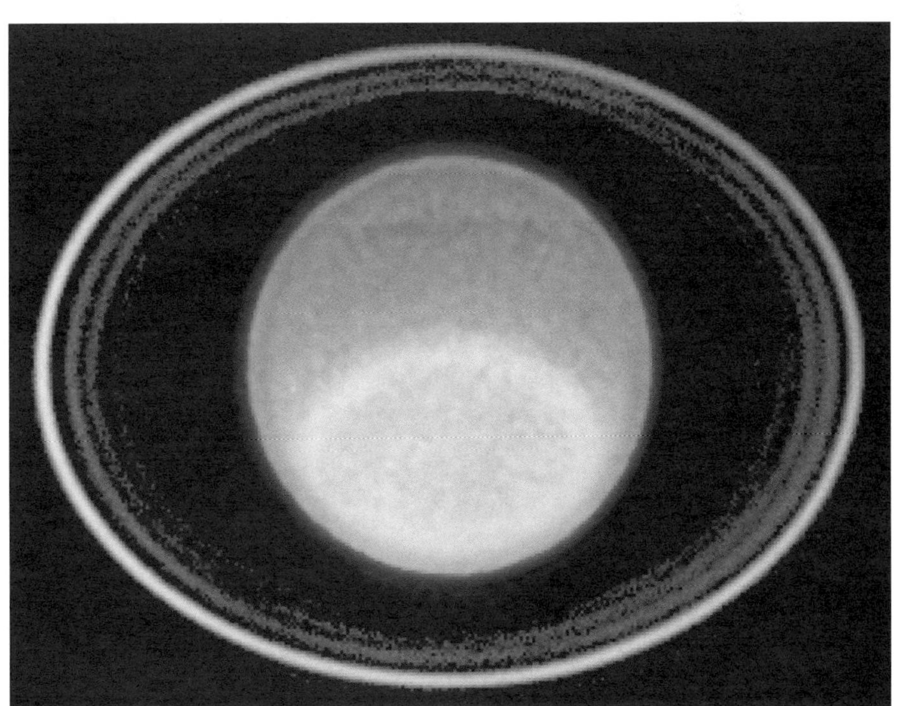

Uranus

PHOTO COURTESY OF NASA

Neptune

Neptune's Properties

Equatorial Radius 24,300 km (= 3.81 R_{earth})

Oblateness 0.026

Mass 1.03 x 10^{29} grams (= 17.2 M_{earth})

Density 1.76 grams/cc

Surface Gravity 1.43 g

Escape Velocity 23.8 km/s

Sidereal Rotation Period 16.05 hours

Solar Day 16.05 hours

Obliquity 29.6°

Sidereal Revolution Period 165 years

Average Distance from the Sun 30.11 AU

Orbital Eccentricity 0.009

Orbital Inclination 1.8°

Number of Moons 8

Following the discovery of Uranus in 1781, astronomers monitored its motion to determine its orbit. After several years, sufficient observations had been made to allow its orbit to be predicted; but in the years that followed, it became apparent that Uranus was not following the predictions, moving too rapidly at first and then too slowly. Several mathematicians and astronomers suspected the presence of another planet, whose gravitational tugs would alter Uranus' orbital speed. Their calculations resulted in the prediction of a new planet beyond Uranus. When the search was finally made in 1846, Neptune was discovered, right where the predictions had placed it.

Neptune is often confused with Uranus because both planets are similar in size and location in the Solar System. Another Jovian planet, Neptune is a bit smaller, slightly more massive, and considerably more colorful than Uranus.

Voyager images from the 1989 flyby show Neptune to be quite blue, due to absorption of red light by methane in its atmosphere. Several interesting features were seen, including small white spots and one large blue oval called the Great Dark Spot. *Voyager* confirmed that Neptune has a set of thin rings and also raised its total of moons from two to eight.

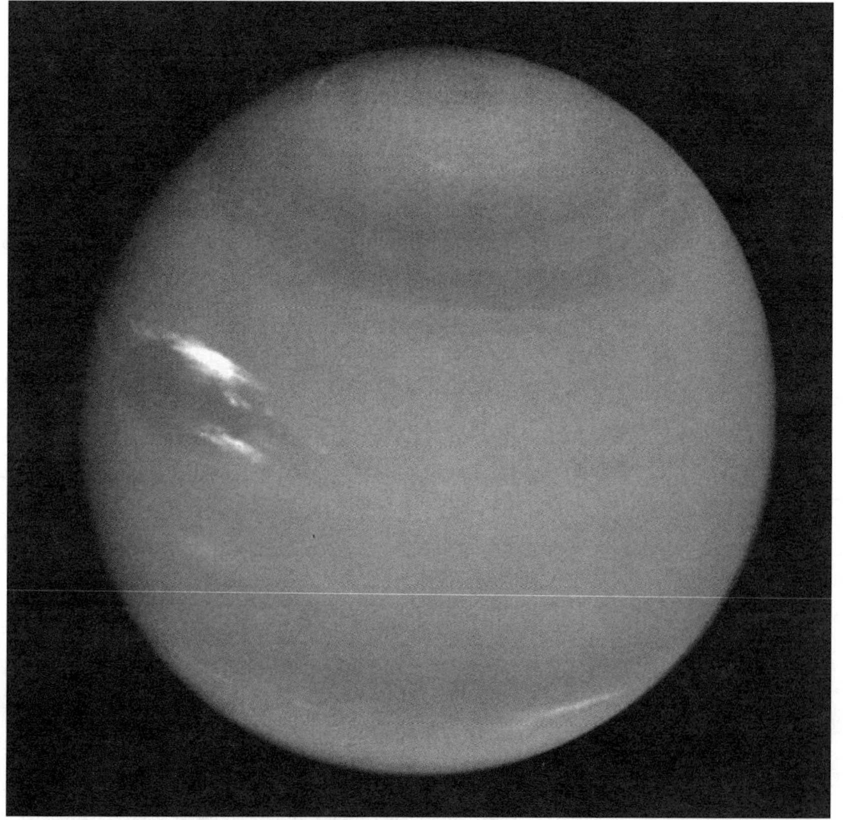

Neptune PHOTO COURTESY OF NASA

Pluto

Following Neptune's discovery, its position was monitored to determine its orbit. Calculation of the orbit followed by further monitoring showed that it was not behaving as predicted. Again, several astronomers made predictions of another planet and began making plans for a search. In 1930, Clyde Tombaugh discovered a new planet, which was named Pluto.

Pluto is a planet of many extremes—its orbit is the largest, most eccentric, and most highly inclined; its average orbital speed is the slowest, and its sidereal orbital period is the longest. Pluto is the smallest, least massive planet—a body even smaller than our Moon. Despite its small size, Pluto has a moon of its own called Charon, whose discovery in 1978 has enabled accurate determinations of Pluto's size and mass.

Pluto's icy composition and frozen atmosphere make it neither a terrestrial nor a Jovian body, and Pluto may have more in common with comets or some of the Jovian moons than with the other planets. Tiny Pluto is far too small to have been responsible for the perturbations in Neptune's orbit, which led astronomers to search for a ninth planet. Much speculation about a tenth planet beyond Pluto has occurred, but none has yet been found.

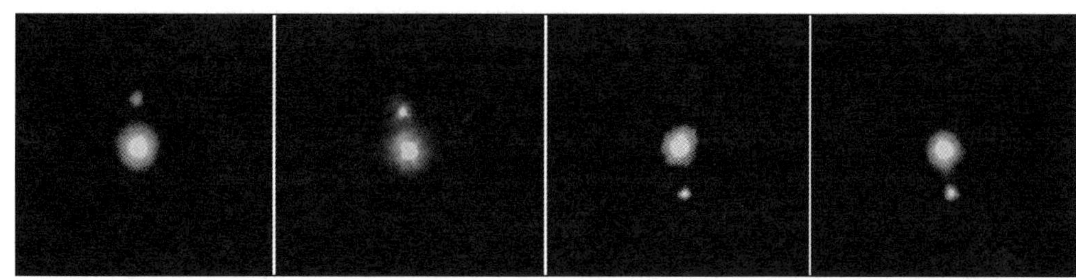

Pluto/Charon

PHOTOS COURTESY OF NOAO

Pluto's Properties

Equatorial Radius 1122 km (= 0.176 R_{earth})

Oblateness 0

Mass 1.5 x 10^{25} grams (= 0.0025 M_{earth})

Density ≈ 2 grams/cc

Surface Gravity 0.03 g ?

Escape Velocity 1.2 km/s ?

Sidereal Rotation Period 6.39 days (retrograde)

Solar Day 6.39 days (retrograde)

Obliquity 62° ?

Sidereal Revolution Period 248 years

Average Distance from the Sun 39.44 AU

Orbital Eccentricity 0.248

Orbital Inclination 17.2°

Number of Moons 1

Asteroids

asteroid
small, rocky chunk of matter orbiting the Sun; also called a *minor planet;* most asteroids orbit in the asteroid belt

asteroid belt
a band between Mars and Jupiter where most asteroids orbit

comet
mass of frozen gases revolving around the Sun, generally in a highly eccentric orbit

minor planet
an asteroid

There are a number of smaller bodies orbiting the Sun; some of these are called **asteroids**, or **minor planets**. Most asteroids seem to have orbits that lie between those of Mars and Jupiter, a region known as the **asteroid belt** (see Figure 7.1). Although there are thousands of asteroids known and probably many more yet undiscovered, the spacing between asteroids is quite large. Space probes to the outer planets have all passed through the asteroid belt without colliding with anything large enough to damage them.

The typical asteroid is not spherical, but irregular in shape, and only a few kilometers across. The largest asteroids are over a hundred kilometers across, but there are not many of these; the total asteroid mass, if combined, would form a body smaller than the Moon. Most asteroids are composed of rocky material believed to be left over from the time of the formation of the Solar System. Whereas most such rocky planetesimals collided with each other and formed into planets, the asteroids were apparently kept stirred up by the tidal forces of Jupiter and never had a chance to form a single body.

Asteroid Gaspra PHOTO COURTESY OF NASA

Asteroids do occasionally collide and break apart, providing fresh debris in the Solar System. Some of these small chunks of rock and iron eventually collide with Earth and produce meteors as they burn up in our atmosphere or meteorites when they land on the Earth's surface. A collision between Earth and a whole asteroid could be disastrous, even if the asteroid were fairly small. Such collisions were common billions of years ago when the Earth was forming, but they are quite rare now. There is some evidence that such a collision might have led to the extinction of the dinosaurs about 65 million years ago.

FIGURE 7.1

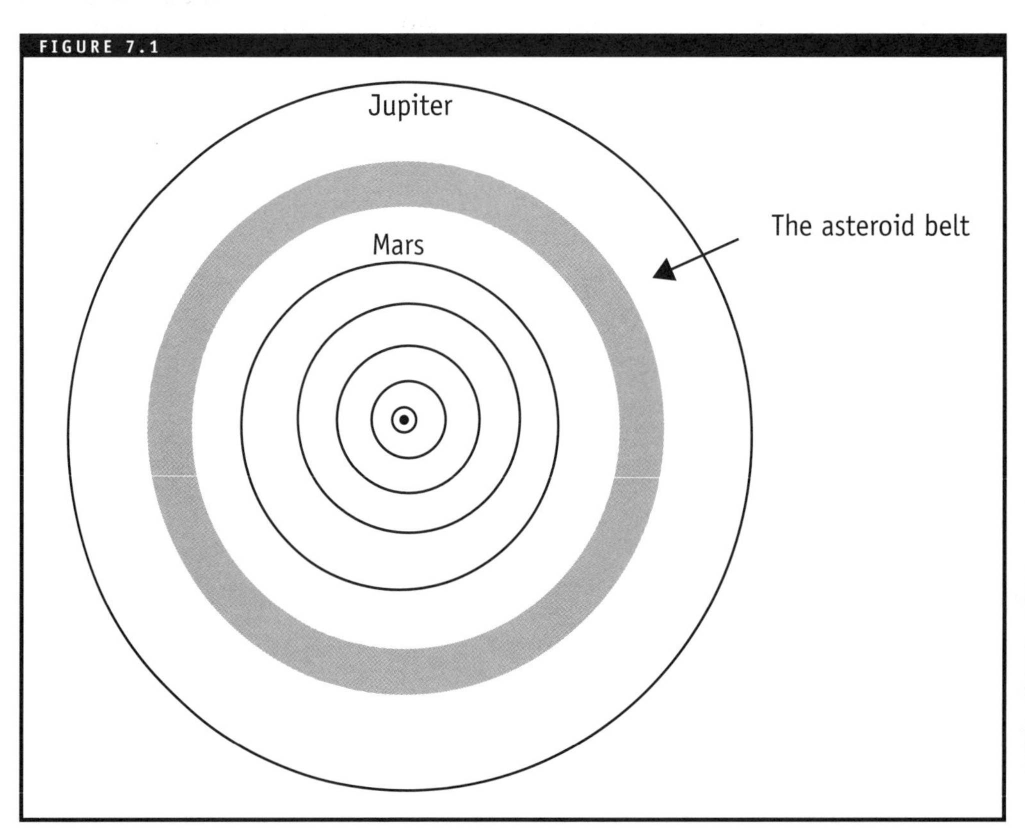

Comets

Other bodies that orbit the Sun are **comets**, which can provide spectacular shows in the night sky. Comets were observed and completely misunderstood by early sky watchers, who thought that these strange lights were part of the Earth's atmosphere. The lights appeared rather suddenly, remained visible for a few weeks, and then disappeared. They did not resemble any of the normal celestial objects—Moon, planets, or stars—being rather fuzzy and often displaying a long tail, as shown in Figure 7.2. They were generally regarded as bad omens, responsible for all sorts of human misery—of which there was plenty—on Earth.

FIGURE 7.2

Two comets with tails

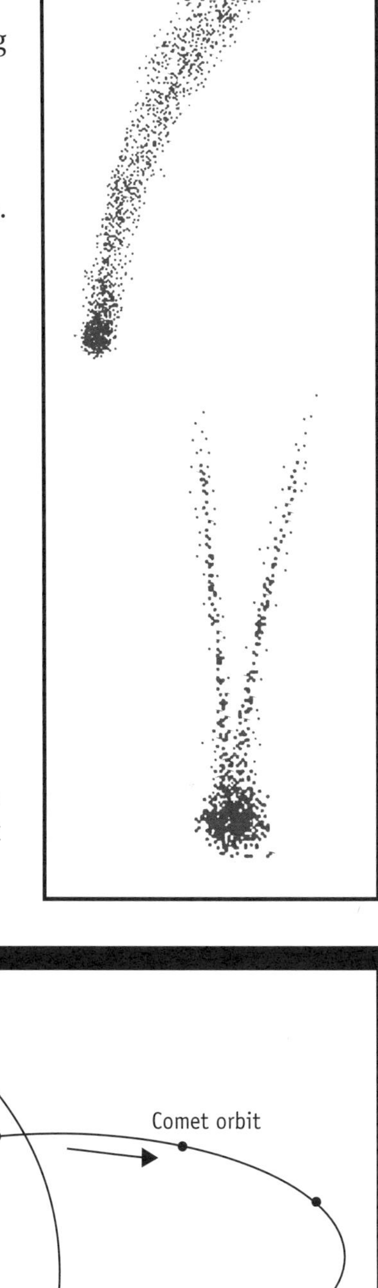

We now know that comets' strange appearance results from their composition. Asteroids tend to be mostly rock; comets seem to be composed of frozen gases of lightweight elements. The standard model of a comet nucleus is a "dirty snowball," a mass of ices a few kilometers across with bits of rock and dust imbedded in it.

These icy planetesimals orbit the Sun in highly elongated orbits. Most of their time is spent far from the Sun, but periodically they head for the inner Solar System, swoop around the Sun, and leave after only a few weeks or months of visibility. A comet's presence is made even more spectacular by the growth of a tail. As the comet nears the Sun, it is warmed by the Sun's radiation. Some of the ice vaporizes and streams out behind the nucleus, forming a tail, which always points away from the Sun, as shown in Figure 7.3. Some of the snowball's dirt is released as the ice melts, providing additional material for the tail and eventually more debris for the Solar System collection.

While there are usually a few comets passing through the inner Solar System at any given time, only rarely do they become bright enough to view without a telescope. Although they can be found in any part of the sky, the brighter ones are often seen best when they are closest to the Sun and have the longest, brightest tails. As such, they will be seen either in the eastern sky before sunrise or in the western sky after sunset (similar to the viewing of Mercury and Venus as morning stars or evening stars, as discussed in Chapter 6). Comets do not streak across the sky; instead, they hang among the stars, gradually changing their position from night to night. Those that pass close to Earth will seem to change most rapidly.

FIGURE 7.3

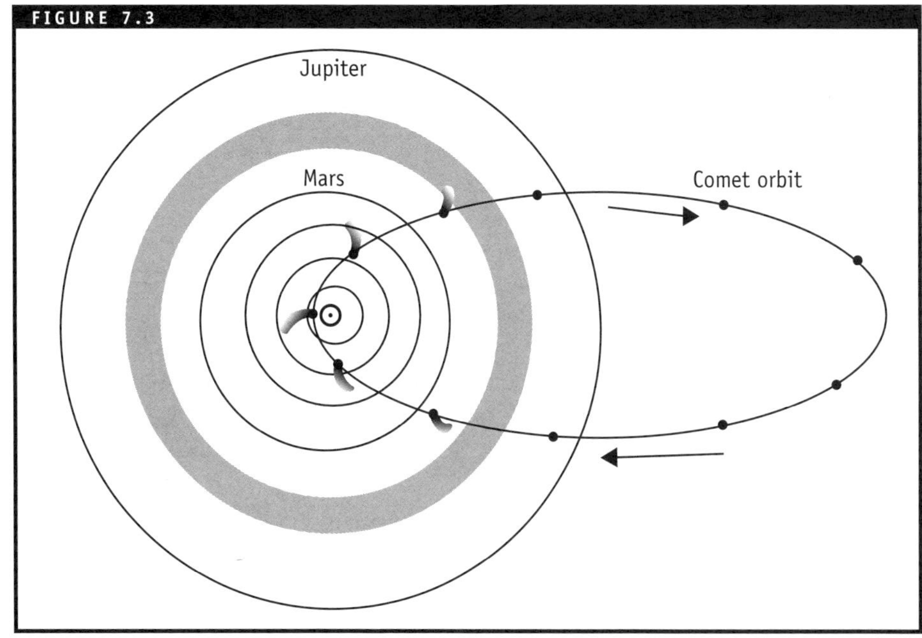

Meteors

meteor
: the luminous trail of heated air produced by a meteoroid's passage through the Earth's atmosphere

meteorite
: a meteroid that reaches the surface of the Earth or another planet or moon

meteoroid
: a rock in space on a collision course with the Earth; sources of meteroids include comets and colliding asteroids

meteor shower
: an event caused by the Earth's passage through the orbit of a comet, where it will collide with an increased number of meteoroids; it happens at the same point in the Earth's orbit—and thus, on the same date every year

radiant
: the point in the sky from which a meteor shower seems to radiate

The Solar System is strewn with small bits of rock that have been produced by collisions between asteroids or released by the evaporation of comets. Bits of space debris that have orbits around the Sun that carry them into the Earth's path are called **meteoroids**. When a meteoroid encounters the Earth, it plunges through the atmosphere at a high rate of speed. The meteoroid's passage through the atmosphere creates friction with the air molecules, heating them enough to cause them to radiate light. The luminous trail of heated air produced in this fashion is called a **meteor**. Other common names for this phenomenon include *shooting star* and *falling star*, but meteors have nothing to do with stars.

Because most meteoroids are quite small, they tend to burn up or vaporize as they pass through the atmosphere. However, the larger ones are not destroyed in this manner but manage to land on the Earth; these rocks are called **meteorites**. Most meteorites are made of stony materials which may be difficult to distinguish from ordinary terrestrial rocks. Some meteorites are made of nickel and iron, and thus are easier to identify as visitors from space.

Meteors are not rare; in fact, they can be seen on any clear, moonless night at a location where the sky is relatively dark. The meteor's streak will be brief, lasting only a second or so before it fades from view. Meteors can occur in any part of the sky at any time, but they are more abundant after midnight when the observer is on the front half of the Earth as it orbits the Sun, plowing through the meteoroids. In the same manner, a car driving down the road on a warm summer evening will collect more bugs on its windshield than on its rear window.

Perseids Meteor Shower August 11, 1999 ©1999 WALLY PACHOLKA

As comets evaporate, they release debris that becomes strewn along their orbits. If the Earth passes through the orbit of a comet, it will collide with an increased number of meteoroids, causing an increased number of meteors. Such an event, called a **meteor shower**, happens at the same point in the Earth's orbit—and thus, on the same date every year. Furthermore, because these meteoroids are all traveling in roughly the same direction in space, the meteors that they create all seem to radiate from the same point in the sky. Meteor showers are named for the constellation in which this **radiant** is located. One of the most famous showers is the Perseids, which occurs around August 11 each year and has a radiant in the constellation Perseus.

Chapter 8
Motions of the Night Sky

The Shane Telescope at Lick Observatory

PHOTO COURTESY OF NASA

The principal celestial bodies discussed in the previous chapters have been a limited number of reasonably bright, close objects—the Sun, Moon, and planets. However, a glance at the night sky reveals many, many more faint points of light—the distant stars. Each star is really a sun, perhaps with planets orbiting around it. Many stars are double or multiple systems with two or more suns orbiting each other. But to us, they appear as small lights spread over the whole sky, in a nonuniform distribution.

Locating particular stars and groups of stars in the night sky is interesting. It is also a practical skill to have if you are ever lost in the woods. Locating stars is not necessarily easy, since the sky appears to move in response to the motions of the Earth. In the next two chapters, you will learn how the Earth's motions affect the appearance of the sky.

The Celestial Sphere

celestial pole
one of two points that mark the intersection of Earth's rotational axis with the celestial sphere; the north celestial pole (NCP) is directly above the North Pole while the south celestial pole (SCP) is directly above the South Pole

celestial sphere
a huge, imaginary sphere around the Earth, used to model the positions and motions of celestial bodies

north celestial pole (NCP)
marks the intersection of Earth's rotational axis with the celestial sphere directly above the North Pole

south celestial pole (SCP)
marks the intersection of Earth's rotational axis with the celestial sphere directly above the South Pole

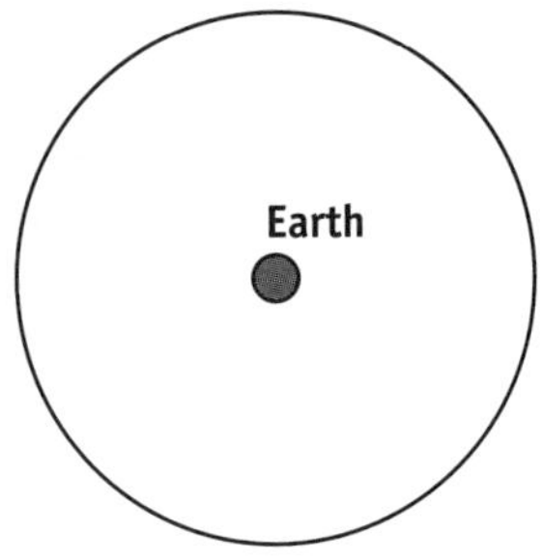

Locating a particular star is easier if you have some means of describing positions in the sky. The first step will be to understand the shape of the sky. Then its apparent motion can be considered.

Space extends in all directions without any boundaries, so discussing its shape may seem odd. In this context, consider the sky as your window to the universe—you look out through the sky to view the stars around you, which are spread throughout space.

Consider a window through which you look to see houses and trees in the distance (Figure 8.1a). Then consider a painting of the same houses and trees, hanging on the wall beside the window (Figure 8.1b).

FIGURE 8.1

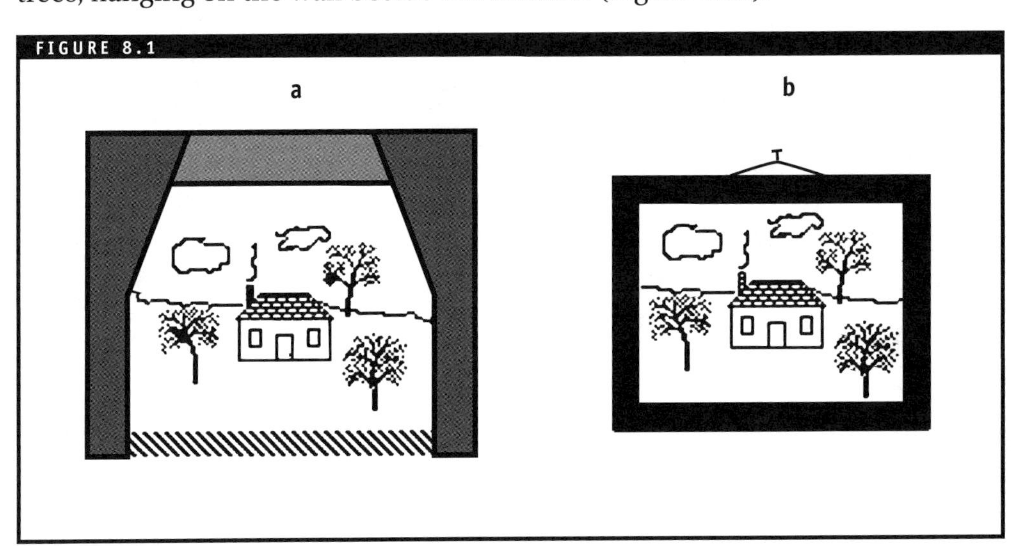

The objects that you see through the window lie at different distances from you, while the objects in the picture are all painted on the same layer of canvas. Obviously, if you are trying to measure the relative distances of the houses and trees, you must use the view through the window. But if you are only concerned with the horizontal and vertical positions of these objects, you could use either the painting or the view to make your measurements, whichever is more convenient. In much the same way, imagine the sky as a surface, with stars glued on it. The stars' actual distances from you are unimportant at this time, and you need not consider them.

When you look up at the sky, you see stars in all directions. It is as if a huge bowl, lined with stars, had been turned upside down over you. Because people on the other side of the Earth will have the same experience, the bowl can be extended to make it a complete sphere around the Earth. This **celestial sphere** is very large—so large that the Earth would be just a tiny speck if it were drawn to scale (see Figure 8.2).

FIGURE 8.2

The celestial sphere

Earth

The celestial sphere is totally fictitious—there is no giant sphere around the Earth. But it is a useful way to describe the sky.

Within the celestial sphere, Earth is not stationary. It rotates much like a globe does—from west to east, or counterclockwise (CCW), as seen looking down on the North Pole. We do not perceive the Earth's rotation; instead, it seems that everything else is moving. Objects in the sky seem to move in a direction opposite to the turning Earth—from east to west. As a result of Earth's rotation, the celestial sphere and the objects on it (the Sun, Moon, stars, and so on) appear to move from east to west, about once per day.

FIGURE 8.3

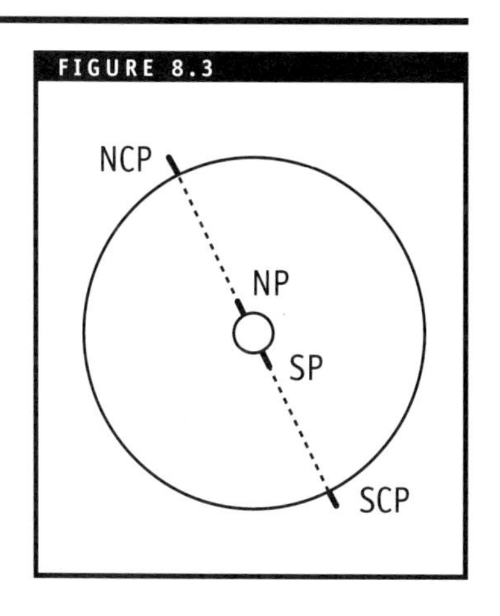

The Earth rotates on an axis through the North and South Poles. We can extend that axis out into space until it intersects the celestial sphere at the north and south celestial poles—the NCP and SCP, as shown in Figure 8.3.

These special points on the celestial sphere, the **celestial poles**, are imaginary—there is no star or other celestial object located there. They serve as reference points in the sky, and are used to locate objects.

Directions in the Sky

On the surface of the Earth, we define north as the direction toward the North Pole. South is toward the South Pole. East is the direction of the Earth's rotation, and west is the opposite.

In a similar manner, you can define directions on the celestial sphere. Just as you can inquire which way is north from a given town, you can ask which way is north from a given star. The answer is similar to that for the Earth—north is the direction toward the **north celestial pole (NCP)**, south is toward the **south celestial pole (SCP)**, west is the direction of the celestial sphere's apparent motion, and east is the opposite. Thus, to someone looking through the celestial sphere at Earth, the directions on each would match up—north with north, east with east, and so on.

However, to an observer on the Earth's surface, located between the Earth and the celestial sphere, the matchup is a bit confusing. Figure 8.4a shows a map of Minnesota with north at the top, west on the left, and so on. Figure 8.4b shows a map of the sky, which has north at the top and west on the right!

The reason for this apparent contradiction is simple. As you look at Minnesota on a globe, you are looking down at the ground, and west is on your left. But in order to look out at the sky, you must turn around, causing west to be to your right. Imagine facing a dance partner; your right hand meets his or her left hand, and you are looking in opposite directions.

FIGURE 8.4

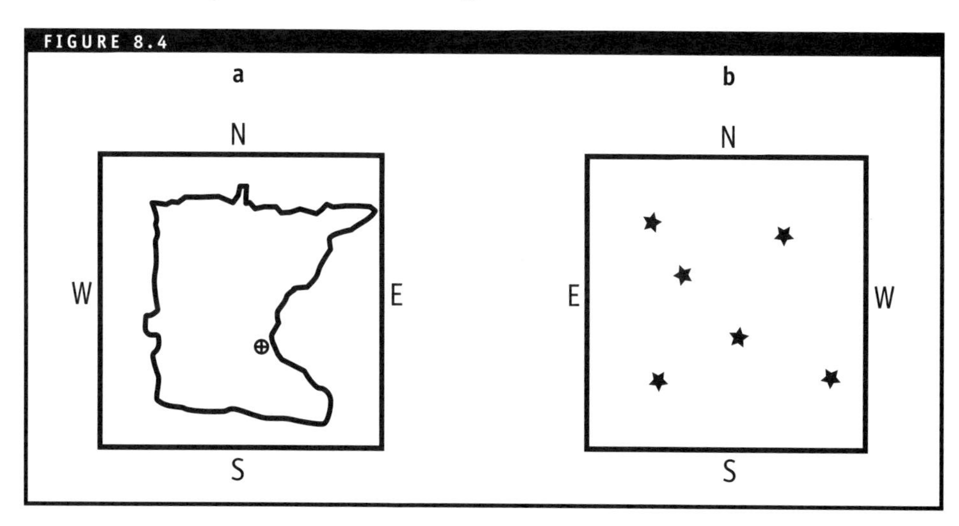

celestial equator
all points on the celestial sphere that are equidistant from both celestial poles

horizon
the circle around you where the sky and the ground meet

nadir
the point in the celestial sphere directly beneath the observer, opposite the zenith

zenith
the point in the sky directly above the observer, opposite the nadir

There is another set of reference points on the Earth, each of which is equidistant from the North and South Poles. They lie in a circle called the equator. There is a similar circle around the celestial sphere, called the **celestial equator** (marked *CE* in Figure 8.5), which lies in the same plane as Earth's equator. Both circles divide their respective spheres into two hemispheres—northern and southern.

Viewing the Sky

If you go outside on a clear night and look up and all around you at the sky, you should be able to see about half of the celestial sphere. You cannot see the other half because the Earth is in the way. Figure 8.6 shows the celestial sphere with a circle dividing it into two halves. This circle, called the **horizon**, divides the sky from the Earth; it separates the part of the celestial sphere that you can see from the part that you can't.

FIGURE 8.5

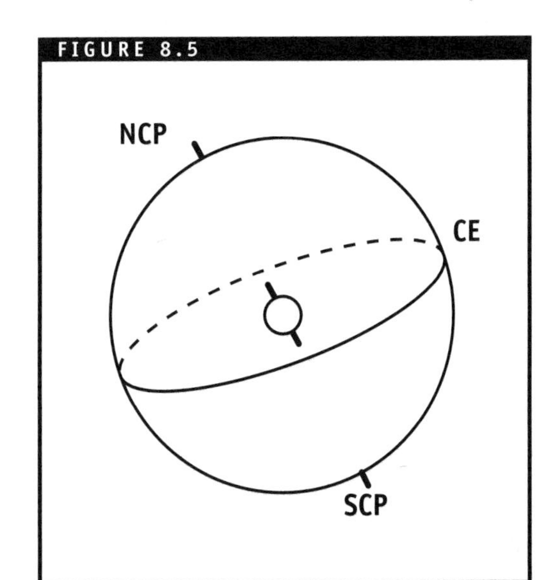

FIGURE 8.6

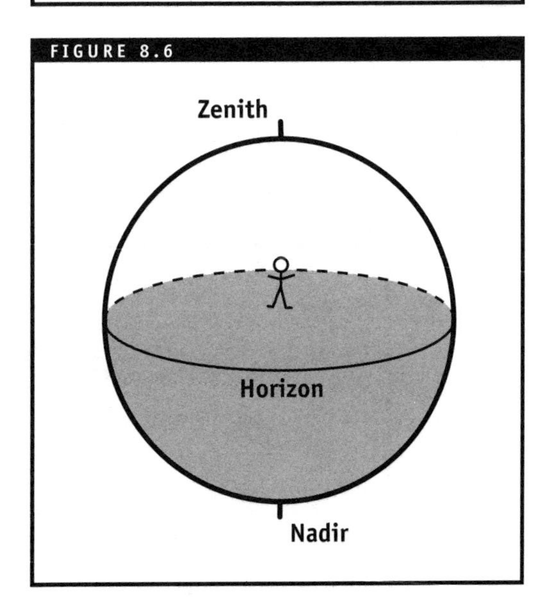

This figure also shows two more special points on the celestial sphere: The point directly overhead is called the **zenith**, while the point directly below (opposite the zenith) is called the **nadir**. The zenith, nadir, and horizon are all fixed to the observer. Your zenith is slightly different from that of your friend seated only a few feet away.

We now have two sets of reference points and circles—how do they match up? As you look at the sky, your zenith is overhead, but where is the north celestial pole? Since the NCP is aligned with the Earth's axis, while the zenith is aligned with you, the relative position of the NCP in your sky will depend on your location on Earth—specifically, on your latitude.

Figure 8.7a shows the view from the North Pole (latitude = 90°). Here the NCP is directly overhead, the SCP is at the observer's nadir, and the celestial equator circles around the horizon. From this location, only the northern celestial hemisphere is above the horizon; the southern celestial hemisphere cannot be seen.

Figure 8.7b shows the view for an observer located at the equator (latitude = 0°). For this person, the celestial equator passes overhead through the zenith and circles down through the nadir, while the celestial poles lie on the northern and southern horizons. Equatorial observers can thus see half of the northern and half of the southern hemispheres at any time.

For middle latitudes, the situation is more complicated. For example, consider the view from Minnesota, located about halfway between the equator and the North Pole. As seen in Figure 8.8, the NCP will be found about halfway between the northern horizon and the zenith, the SCP will lie below the southern horizon, about halfway from the horizon to the nadir,

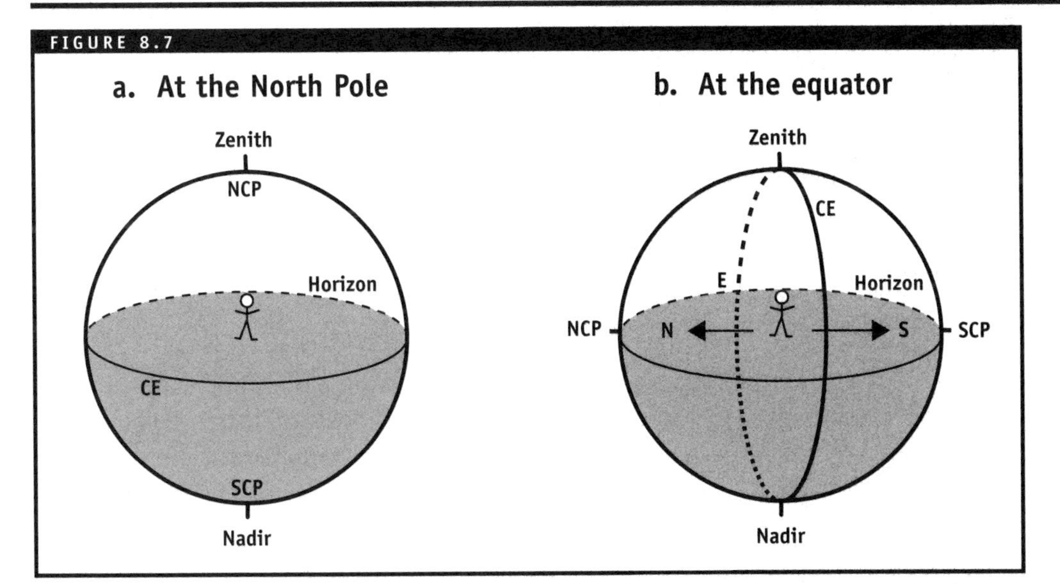

and the celestial equator will intersect the horizon at an angle, reaching halfway up to the zenith and halfway down to the nadir. Most of the sky visible to Minnesotans is in the northern celestial hemisphere, but a small portion of the southern hemisphere is also seen.

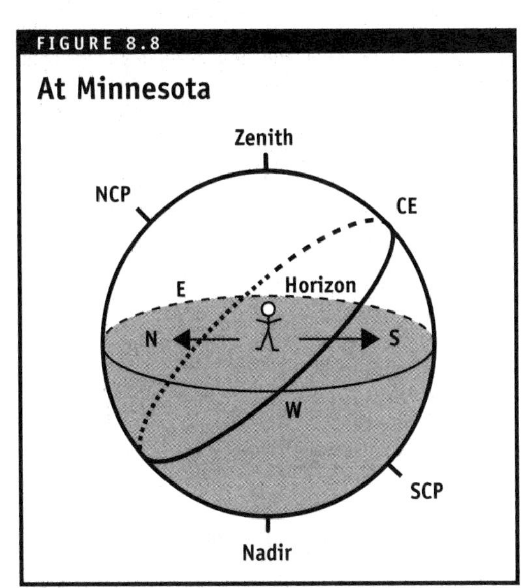

The precise positions and angles depend on the observer's latitude. For an observer at Mankato, Minnesota (latitude = 44° N), the NCP is 44° above the northern horizon and the SCP is 44° below the southern horizon. The celestial equator reaches 46° above the southern horizon, dips 46° below the northern horizon, and makes an angle of 46° with the horizon where it crosses. An observer in Austin, Texas (latitude = 30°) would have the NCP 30° above the northern horizon and the SCP 30° below the southern horizon; the celestial equator would reach from 60° above the southern horizon to 60° below the northern horizon, making an angle of 60° where it crosses the horizon.

Effects of Rotation

Your latitude fixes the position of the north celestial pole in your sky; Earth's rotation causes the stars to appear to circle around it. Rotation of the Earth gradually varies the direction in which you are looking—much as a rider on a merry-go-round has a constantly changing view of the rest of the fair. Just as the rides, booths, and people at the fair seem to fly by as you ride the merry-go-round, the stars will appear to move across the sky as the Earth turns, although much more slowly. The paths that the stars follow across the sky depend on their locations on the celestial sphere. The direction in which they move is simple—as the Earth rotates eastward, the celestial sphere appears to move westward. Thus, the stars constantly move westward across the sky (with west as indicated in Figure 8.4b).

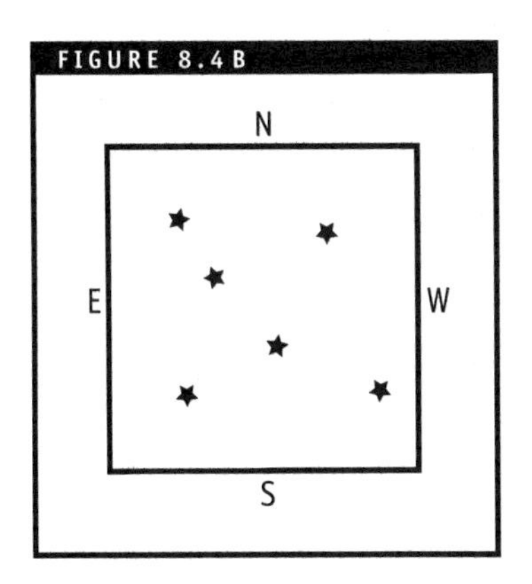

diurnal motion
daily motions of the Earth and sky, associated with the rotation of the Earth

star trails
streaks made by stars on photographs of the sky, caused by Earth's rotation

The simplest way to picture the motion of points on the rotating celestial sphere is to observe another rotating sphere, such as a spinning globe. As the globe spins, the North Pole remains in the same place; similarly, the NCP does not move but stays in the same place in our sky. Points on the globe close to the North Pole (such as the northern tip of Greenland) move in small circles around the pole as the globe spins, while points farther from the pole (such as the British Isles) move in larger circles. In the same manner, stars near the NCP move in small circles around it, while stars farther away move in even larger circles.

When looking directly at the equator on the spinning globe, you will see that although it moves in a circle, its motion appears straight. Similarly, the stars at the celestial equator seem to follow straight paths rather than circular ones. Since we seem to be at the center of the circle they follow, we do not readily perceive the curvature of their paths.

We have a similar problem when looking at the horizon, which extends as a huge circle around us. Despite its circular nature, any small portion of the horizon appears as a straight line dividing the sky from the ground (ignoring features such as mountains or hills).

Of course, the motion of the sky is gradual because the Earth turns slowly; therefore, the movement of the stars is not immediately apparent to the casual observer. Over a sufficient length of time, it might be noted that a given star has changed its position (with respect to a reference point on the horizon, such as a tree or building), but this observation requires that the observer remain in the same position over this interval in order to notice the effect.

In order to demonstrate that this motion takes place, we need either a long time frame—so that changes will be quite obvious—or a device that can continuously record positional changes. A camera equipped with a *B* or *Bulb* setting on the shutter is a good device for this purpose. This setting allows the shutter to be locked open for any length of time. Pointing such a camera at the sky on a clear, moonless night, away from streetlights, and locking the shutter open for a period of several minutes will produce a picture showing the **diurnal motion** of the stars. The streaks across the film left by the stars as they move are called **star trails**. Longer exposures produce longer trails.

Star trails PHOTO COURTESY OF NOAO

Star Trails

The shape of star trails depends on which part of the celestial sphere is photographed, the observer's latitude, and the direction in which the camera is pointed. To illustrate, take a "trip" to different locations on the Earth—the same places visited above.

Your first stop is the North Pole, where the sky appears as shown in Figure 8.7a. At this location, point your camera in two different directions—straight up and toward the southern horizon.

If you point a camera straight up at the zenith, the NCP will be in the center of your picture. Since the stars appear to move in circles around the celestial poles, you would photograph circular star trails. The direction of the stars' motion will be westward, or counterclockwise around the NCP, as indicated in Figure 8.9a. Figure 8.4b shows that on a star map oriented to north at the top, west is to the right. For a star at the bottom of Figure 8.9a, north is up (toward the center of the circles) and therefore west would be to the right. This results in counterclockwise motion around the NCP.

From the North Pole, the celestial equator lies right along the horizon. When you look south from the North Pole, every direction is south and stars will move parallel to the horizon from east to west (left to right); these star trails will be nearly straight lines, as in Figure 8.9b.

At the North Pole, the stars circle counterclockwise around the NCP and cruise parallel to the horizon from left to right (east to west), never rising or setting. The same stars—those in the northern celestial hemisphere—are always above the horizon, while the other half of the celestial sphere remains hidden from view below the horizon.

If you were to go to the South Pole, you should find a similar view, but with directions reversed. The stars would circle clockwise around the SCP at the zenith and cruise parallel to the horizon from right to left (but still east to west), again never rising or setting. Only southern celestial hemisphere stars would be visible, with the northern celestial hemisphere always below the horizon.

FIGURE 8.9

Star trails at the North Pole

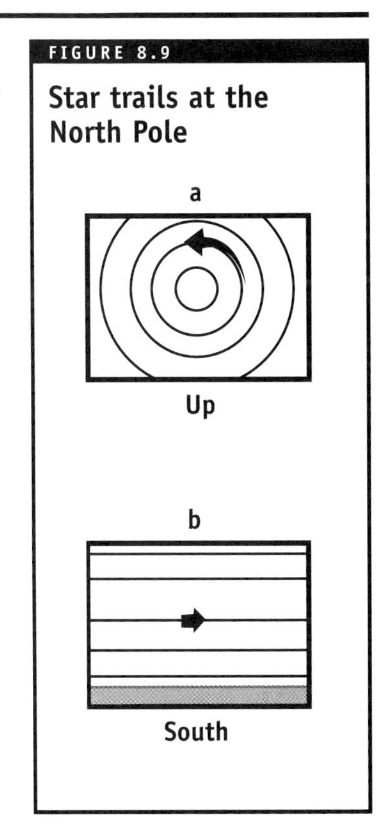

FIGURE 8.4 B

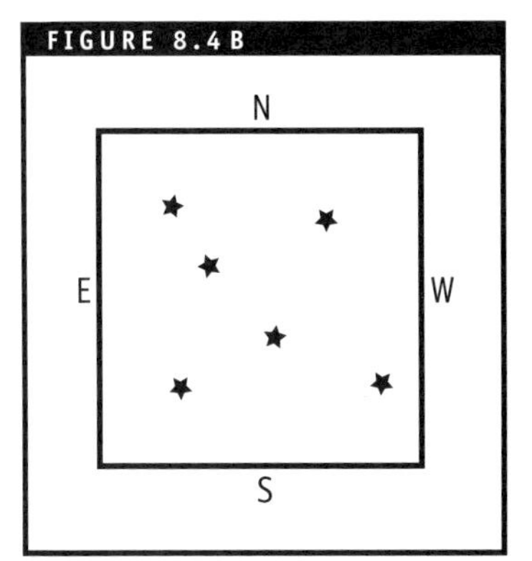

FIGURE 8.7

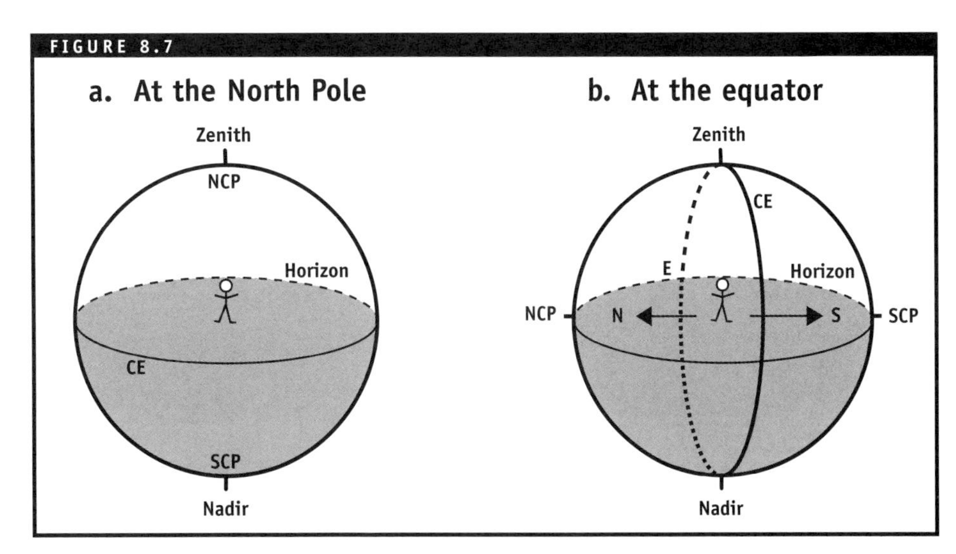

FIGURE 8.10

Star trails at the equator

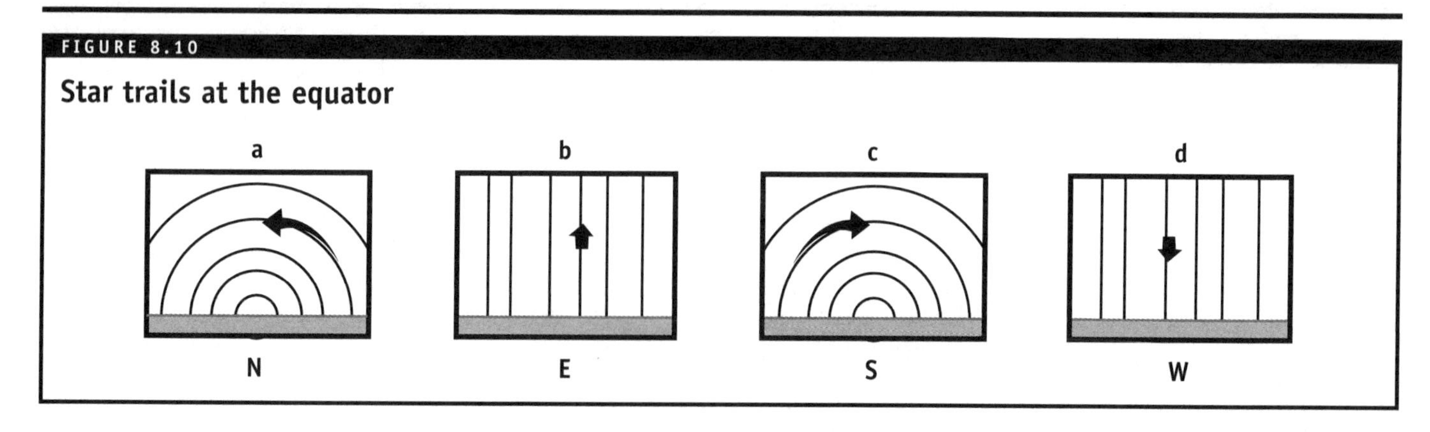

Wherever you go, certain constants will remain—the stars will always move counterclockwise around the **NCP**, clockwise around the **SCP**, and parallel to the **celestial equator** from east to west.

The next stop is the equator. As Figure 8.7b shows, the NCP and SCP will be on your northern and southern horizons, respectively. Therefore, when you look north, you should see stars moving in counterclockwise circles around the NCP (Figure 8.10a) while to the south, they will move clockwise around the SCP (Figure 8.10c). In each case, star trails will be semicircles. To the east or west of the equatorial observer, the celestial equator intersects the horizon at a 90° angle. Star trails in these directions will be vertical, with the stars moving upward in the east (Figure 8.10b) and downward in the west (Figure 8.10d).

Of course, relatively few people live at the poles or the equator; most prefer the climate at more moderate latitudes. For a representative view, revisit Mankato, Minnesota. Here the NCP is located 44° above the northern horizon, and the SCP is hidden below the southern horizon, as in Figure 8.8. Pointing your camera to the north, you will find the NCP at the top of, or just above, your picture, and the stars moving in counterclockwise circles around it. Only the lower portions of these circles appear in your picture, as concave-upward arcs (Figure 8.11a). With the camera pointed south, you can photograph only the upper portions of the clockwise circles around the SCP, producing concave-downward trails (Figure 8.11c).

FIGURE 8.7 B

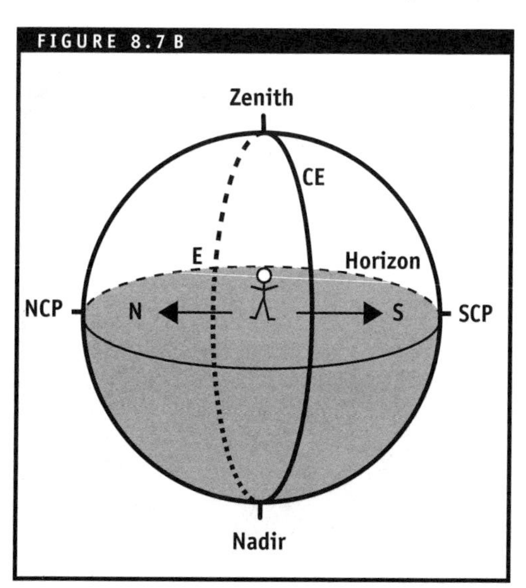

FIGURE 8.8

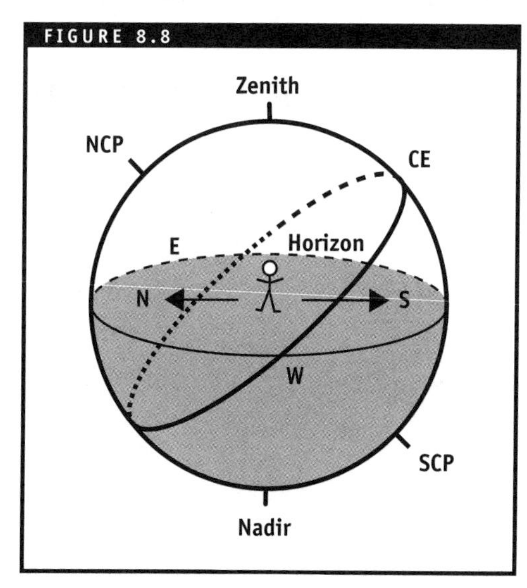

Figure 8.8 shows that the celestial equator intersects the horizon at slants in the east and west. These slants show up in the star trails for these directions—the stars move up and to the right as they rise along the eastern horizon (Figure 8.11b) and down and to the right as they set along the western horizon (Figure 8.11d).

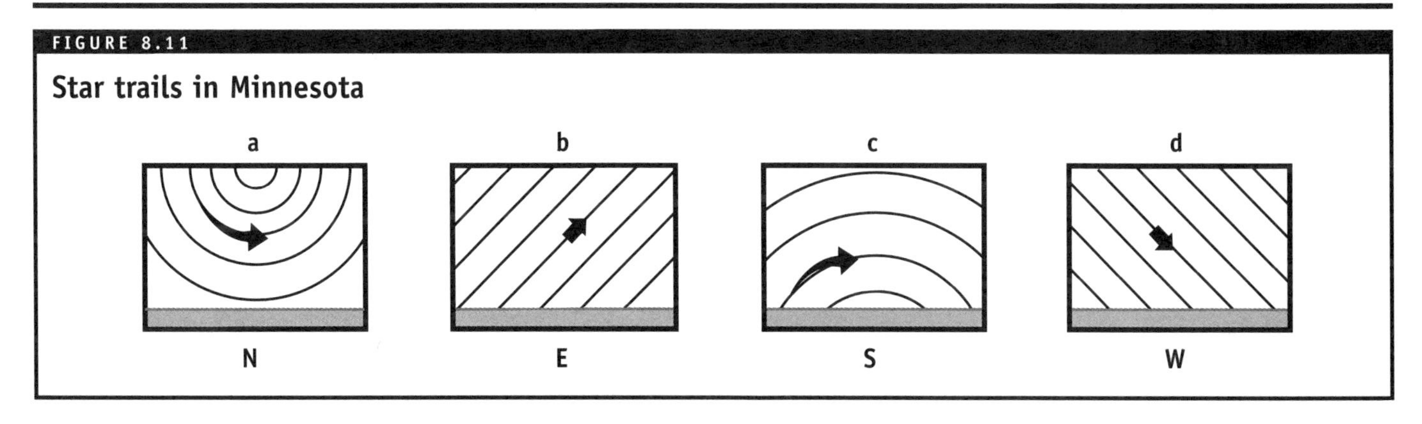

Starrise and Starset

In general, stars rise along the eastern horizon, move across the sky as shown by the star trails, and set along the western horizon. A star that rises directly east will set directly west. One that rises south of east will set south of west; and one that rises north of east will set north of west.

Depending on the observer's latitude, there may be some stars that never rise or set; their star trails do not intersect the horizon. Such stars are called **circumpolar stars** because they circle near the celestial poles. At the North Pole, all stars are circumpolar, while at the equator, none are. At other latitudes there will be some circumpolar stars and others that rise and set regularly. Figure 8.12 shows the location of the circumpolar stars for observers in Mankato, Minnesota.

Stars in the shaded region around the NCP are circumpolar because they never go below the horizon. Stars in the shaded region around the SCP are also circumpolar, but they can never be seen from this latitude because the Earth's rotation never carries them above the horizon. The shaded regions of circumpolar stars would be smaller for observers at latitudes closer to the equator.

celestial equator
all points on the celestial sphere that are equidistant from both celestial poles

circumpolar stars
stars that never rise or set because their star trails do not intersect the horizon

NCP
north celestial pole

SCP
south celestial pole

In summary, the Earth's eastward rotation causes the sky to appear to turn from east to west every day. Because the Earth is spherical, our view of the sky varies with latitude.

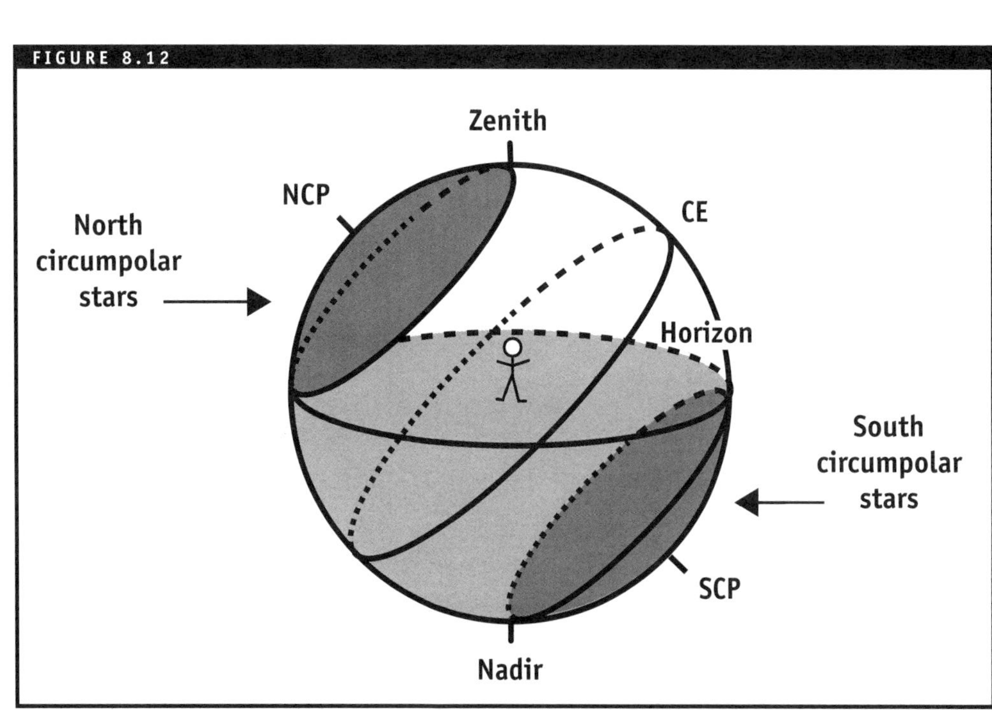

ecliptic
the Sun's apparent path on the celestial sphere; also, the plane of the Earth's orbit; the ecliptic is not in the same plane as the celestial equator, but inclined to it by about $23\frac{1}{2}$ degrees; this angle is called the obliquity, or the tilt

obliquity (tilt)
the angle between a planet's rotational axis and its orbital axis

FIGURE 8.13

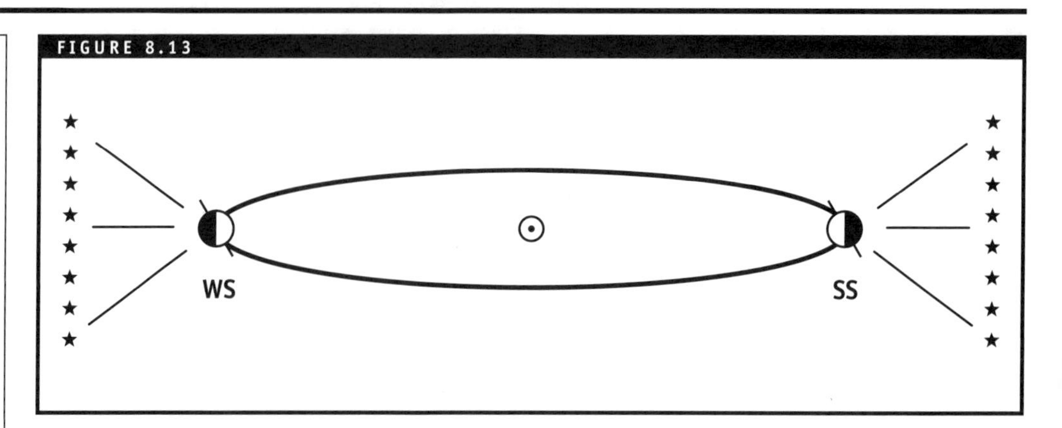

Effects of Revolution

The Earth's revolution also produces changes in the night sky. The rotation period is about one day; the revolution period is about one year. Changes in the appearance of the sky due to revolution will be more gradual than the diurnal changes caused by rotation. To illustrate why revolution changes our view of the sky, consider the following example.

Figure 8.13 shows the Earth at two different positions in its orbit around the Sun. At the winter solstice position, a person observing the night sky would see stars as shown on the left side of the celestial sphere. The same person observing the night sky six months later at the summer solstice position would see stars on the opposite side of the celestial sphere. Different stars are seen in different seasons as the night side of the Earth is directed toward different portions of the celestial sphere.

The rate of change is very gradual. From one night to the next, not much difference would be seen; but over the course of several weeks, the accumulated change becomes quite evident. Thus, both rotation and revolution cause changes in the night sky, but at different rates. The actual position of the stars at any given moment depends on the time of day or night (rotation position) and the day of the year (revolution position).

Another way to think about the orbital changes is to consider the Sun's position among the stars. The Sun's brilliance prevents us from seeing stars in the daytime sky, but they are out there just the same. As the Earth orbits the Sun, the Sun appears to change its position among the stars, as shown in Figure 8.14a. The Earth's counterclockwise motion around the Sun causes the Sun to appear to move counterclockwise around the Earth, or eastward, with respect to the stars, as shown in Figure 8.14b.

FIGURE 8.14

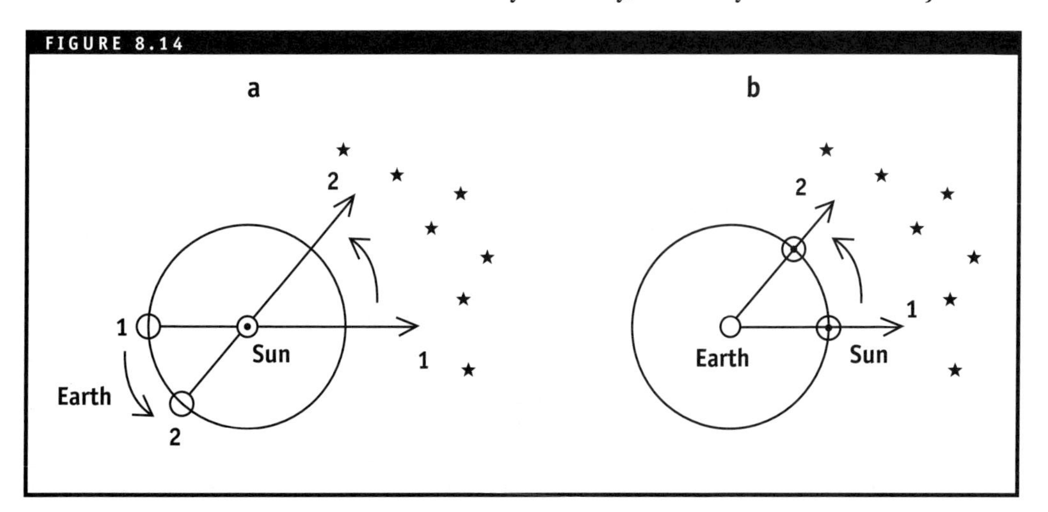

In the fixed-Earth view in Figure 8.14b, the Sun appears to travel in a circular path around the Earth. This path is represented as a circle around the celestial sphere; it is called the **ecliptic**. As seen in Figure 8.15, the ecliptic is not in the same plane as the celestial equator, but inclined to it by about 23½ degrees. This angle is called the **obliquity**, or the tilt.

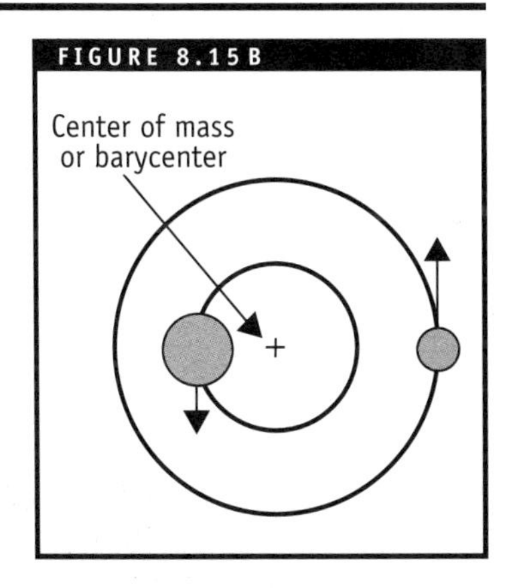
FIGURE 8.15 B

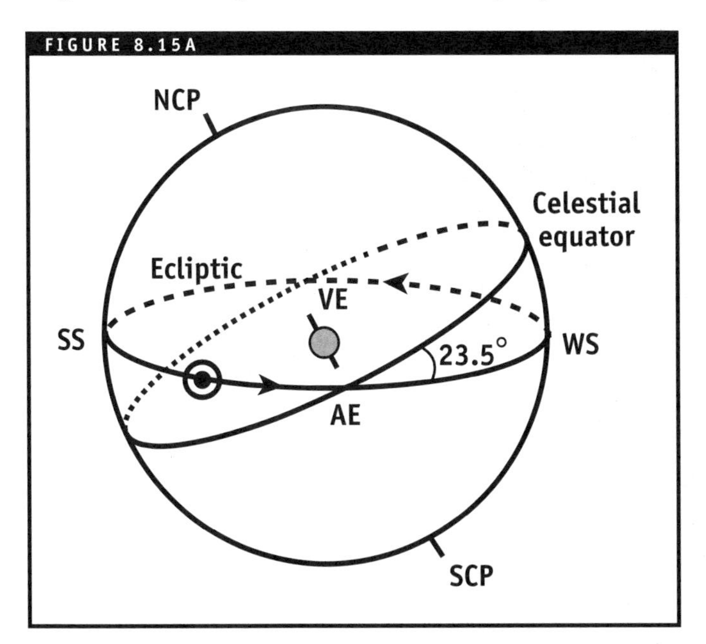
FIGURE 8.15 A

The ecliptic and the celestial equator do not coincide because they are related to separate motions of the Earth. The celestial equator is determined by the Earth's rotation, while the ecliptic is determined by the Earth's revolution. As these two motions are independent of each other, there is no reason for the ecliptic and the celestial equator to align.

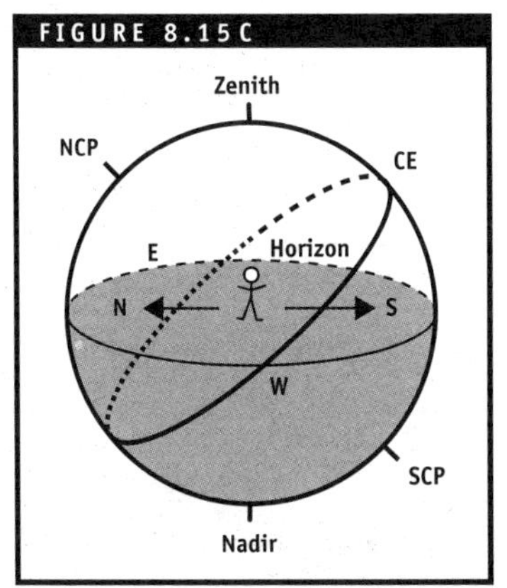
FIGURE 8.15 C

The Sun's motion in Figure 8.15a is counterclockwise around the ecliptic, from the summer solstice (SS) to the autumnal equinox (AE), the winter solstice (WS), and the vernal equinox (VE). Note that when the Sun is at the summer solstice, the North Pole is tilted toward the Sun, while at the winter solstice, the North Pole is tilted away from the Sun. Compare this fixed-Earth view with the fixed-Sun view of Figure 8.15b.

The Sun's motion along the ecliptic explains the variation in sunrise and sunset directions discussed in Chapter 3. When the Sun is on the portion of the ecliptic that lies in the northern celestial hemisphere, it will be north of the celestial equator. Because the celestial equator rises directly east and sets directly west (see Figure 8.15c), the Sun (which is north of the celestial equator) must rise north of east and set north of west during this period, as shown in Figure 8.16. This occurs from the vernal equinox, through the summer solstice, and to the autumnal equinox—the seasons that we call spring and summer. Similarly, during autumn and winter, the Sun travels along the southern portion of the ecliptic, staying south of the celestial equator; and therefore, must rise south of east and set south of west. Only at the equinoxes does the Sun rise directly east and set directly west.

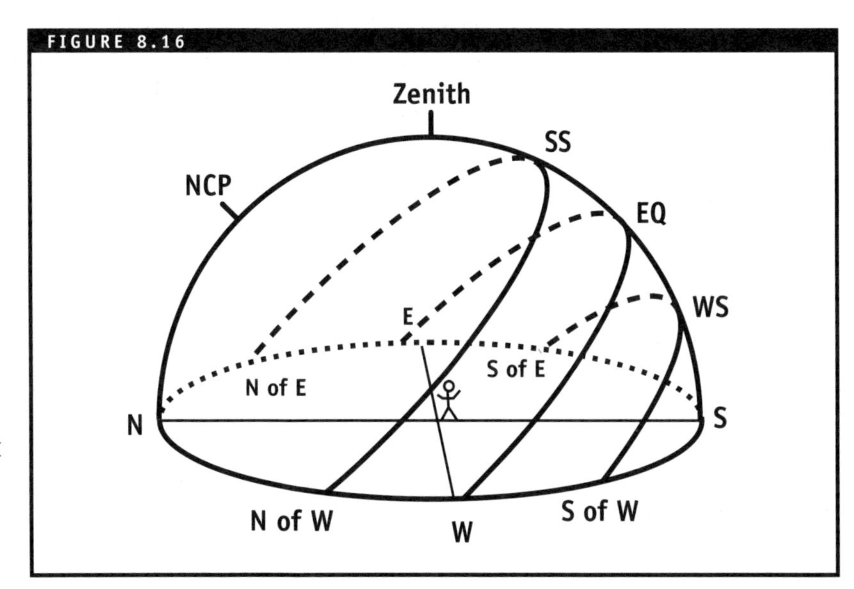
FIGURE 8.16

Chapter 9
Constellations and Star Charts

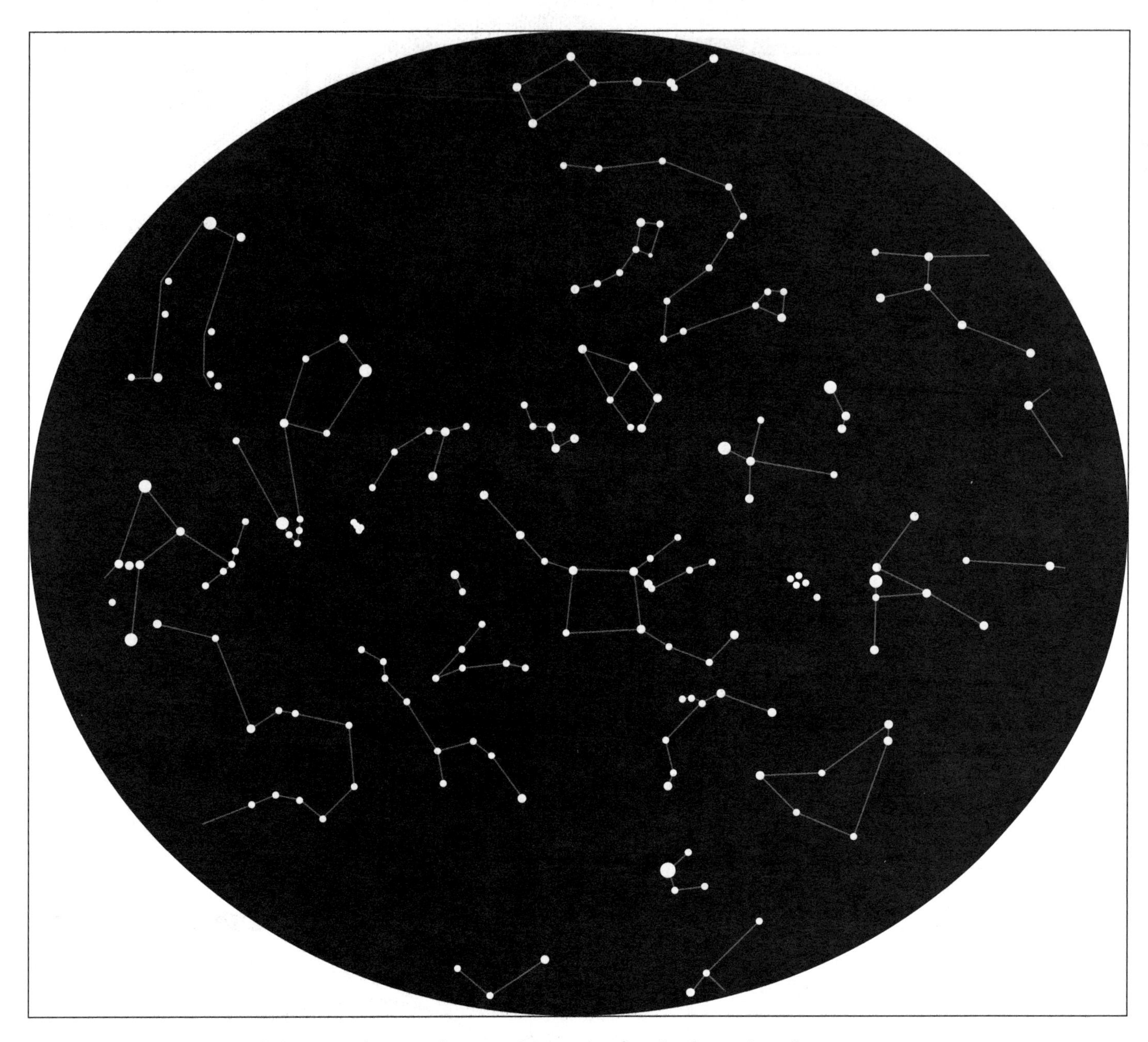

In this chapter you will discover how to locate objects in the sky by using the concept of the **celestial sphere**.The first objects to consider are the stars, of which there are a great many. The stars are all so far away that although they are moving through space at rates comparable to the Earth's orbital speed around the Sun, to us they appear to be fixed in position. For this reason, we often refer to them as **fixed stars**; they occupy fixed points on the celestial sphere.

With very careful measurements of stellar positions made over many years, the motions of some stars can be detected. The problem is similar to measuring the growth of a tree from 50 miles away.

Constellations

celestial sphere
: a huge, imaginary sphere around the Earth, used to model the positions and motions of celestial bodies

constellation
: a group of stars; also, a region of the celestial sphere

fixed star
: a star that occupies a fixed point on the celestial sphere

star chart
: a map of the celestial sphere showing the locations of fixed stars

Throughout human history, skywatchers have grouped and named the stars to make it easier to talk about or study them. The most important function of astronomy in the earlier stages of human development was as the basis for a calendar to show the passage of seasons. Naming the stars made it easier to talk about them. Since there are so many stars, it was convenient to group them together into easily recognized patterns in the sky. These sky groups are what we call **constellations**. They may be large or small and contain stars which are bright or faint. They have all been devised by human observers on Earth, with different cultures inventing different constellations for the same stars. The Egyptians, Chinese, and Maya all had astronomers, and astronomy played an important part in their development.

We have inherited the constellations devised by the ancient Greeks. The Greek culture was dominant in its day, and its astronomical traditions were carried on by Arab astronomers during the Dark Ages in Europe. During the Renaissance, the Greek and Arabic influences on astronomy were absorbed and preserved by successive civilizations, providing the basis for astronomical knowledge today.

Most constellations are modeled after human figures, animals, or other beasts from Greek mythology. In many cases, the brighter stars of a constellation trace an outline or stick figure of the person or animal portrayed—a particular star might be an eye or a head; a group of stars may be an arm, leg, or tail. In nearly every case, a great deal of imagination is needed to see the person, animal, or object that the constellation represents.

The Greeks did not cover the entire sky with constellations, for they could not see all of the celestial sphere from Greece. In addition, the constellations they made did not fit neatly together like puzzle pieces, covering all of their visible sky. The spaces of sky in between that were not in any particular constellation were said to be "unformed." Because these regions did not contain any significant bright stars, this was not a problem for the Greeks.

With the invention of the telescope in 1609 came the revelation that many of the starless "unformed" regions between the constellations were actually populated by numerous faint stars. To accommodate this discovery, the constellation "skeletons" filled out to acquire boundaries, expanding until they touched each other. In some cases, new constellations were invented to fill in spaces between the old ones. And as European explorers circled the globe, they labeled the extreme southern skies with additional constellations to aid in their navigation of southern seas.

By the start of the 20th century, the celestial sphere was covered by over 100 constellations with poorly defined borders. The situation was clarified in 1928 when the number of constellations was reduced to 88, with borders drawn as straight lines running north-south and east-west. The sky that we observe is essentially the same as it was for the Greeks. The difference is in the star charts that we use to identify the constellations.

Star Charts

A **star chart** is a map of the celestial sphere; it shows the locations of fixed stars, which do not change position significantly. Objects that are less stationary, such as the Sun, Moon, and planets, are generally not shown on star charts unless the chart is intended for use over a limited time interval, such as one month. An individual star chart usually covers only a portion of the celestial sphere; a whole set of star charts, numbering anywhere from two or three to over a hundred, is needed to map the entire sky. There are many good sky maps on the market with different layouts.

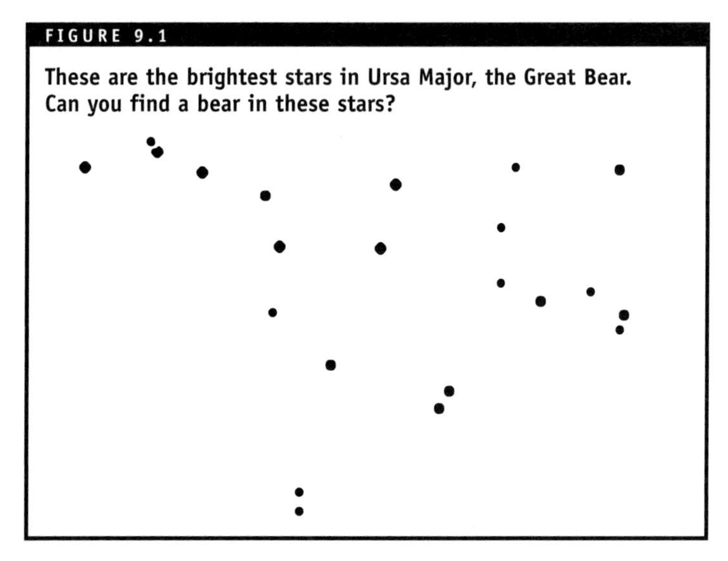

FIGURE 9.1

These are the brightest stars in Ursa Major, the Great Bear. Can you find a bear in these stars?

The appearance of the night sky changes continually due to the Earth's rotation and revolution. Additionally, the appearance of the sky depends on the observer's latitude. One type of star chart shows the night sky as it appears from a particular latitude on a particular date at a particular time of night. Since most people who observe the sky do so in the hours just after sunset, such charts are usually set for this time of night. Because most of the United States lies in a relatively narrow band of latitude (about 25° to 45°), many charts are drawn for an average value of 35° to 40°, showing a sky that is fairly accurate for most of us. This leaves only the date as a major variable. Complete sky maps are often produced in sets of 12 monthly star charts, like those found in issues of *Sky and Telescope* or *Astronomy*. One need only select the chart for the current month in order to begin to identify constellations.

These charts are fine to use as long as one only makes observations at the times for which the charts are drawn. However, an observer going out at 2 a.m. in October will find that the October star chart drawn for 8 p.m. is not a very good match for the sky that he or she sees. The situation can be easily remedied—for every two hours past the recommended observing time, add an extra month. Since 2 a.m. is six hours later than 8 p.m., the star chart for January at 8 p.m. (three months past October) will work fine.

The other method of making sky maps ignores variations due to latitude, date, and time. The celestial sphere is divided into regions, and a separate star chart is drawn for each one. For example, a chart is often made of the region around the north celestial pole, showing the north circumpolar constellations. If you are outside looking at this part of the celestial sphere, you would use that particular chart. The trick is that you need to know at which part of the sky you are looking in order to know which chart to use. Generally, the more charts in a sky map, the smaller the area of sky covered, and the greater the detail shown on each chart.

To choose the type of sky map you want to use, think about map depictions of the United States. We could show the United States on one map, with principal cities and highways marked; or we could show the same area on

50 state maps, with all towns, rivers, and paved roads marked; or we could collect a set of several hundred county maps from all the states, showing all the roads, streams, bridges, and so on. You must decide how much detail you need and choose your map accordingly.

Most star charts will show the stars as dots of different sizes, possibly adorned with small points or rays. The different sizes or points indicate the brightness of the stars; bigger dots indicate brighter stars. If you hold the chart at arm's length, you should be able to see the biggest dots (brightest stars) most easily. Identifying constellations is a matter of matching stars in the sky to stars on the chart. If you are correct, the patterns should be the same. Some charts will also have a few lines connecting the stars within a constellation. These lines will not show up in the sky, but they may help you see the bear, dog, or person that the constellation is supposed to be. An example of a familiar constellation is shown in Figure 9.1.

Where in the sky are constellations found? As mentioned above, the stars are fixed. That means they do not move around on the celestial sphere. But as both rotation and revolution cause changes in our view of the celestial sphere, when we look to the east, we are not always looking at the same constellations.

A constellation near the north celestial pole (such as Ursa Major) will remain near it as the Earth rotates and revolves. To find such a constellation, look in the general direction of the north celestial pole and try to match the stars to those on a chart of the north circumpolar region. You may have to rotate the chart to match it with the sky.

A constellation located along the celestial equator (such as Orion) will rise in the east, move across the southern sky, and set in the west (for a northern hemisphere observer). Because of its motion, there is no single, simple direction that can be given to find Orion in the sky—in fact, half the time it is not even above the horizon. Orion is most easily found by noting its distinctive pattern of stars (see Figure 9.2) and watching for it in the part of the sky where the celestial equator lies at your latitude. It also helps to know which time of year Orion should be visible in the evening sky (December through April).

FIGURE 9.2

These are the brightest stars in Orion, the Hunter. Can you find a hunter in these stars?

Star Names

Individual stars can be identified fairly easily once constellations are found. The brighter stars usually have proper names, such as Sirius, Pollux, or Antares. Many stars are also assigned Greek letters as names (with a form of the Latin name of the constellation added): Sirius (in Canis Major) is α Canis Majoris; Pollux (in Gemini) is ß Geminorum; Antares (in Scorpius) is α Scorpii. Knowing a few stars by name makes it easier to identify the constellations; it also makes it easier to find the planets, which often masquerade as bright stars.

The Zodiac

The Moon and planets move around the sky in an orderly, predictable fashion, passing stars as they go. We have seen that the Sun has a similar motion, moving around the **ecliptic** each year. The Moon's orbit is inclined by 5° to the Sun's orbit; thus, the Moon never strays very far from the ecliptic in its monthly journey around the sky. The Solar System is somewhat disk-shaped, with the planets' orbital planes all lying fairly close to the plane of the ecliptic. This keeps the planets near the ecliptic as they travel around the sky. The region of sky through which the Sun, Moon, and planets move is called the zodiac. The zodiac is a band around the celestial sphere—18° wide and centered on the ecliptic, as shown in Figure 9.3.

Locating planets can often be as simple as determining in which zodiacal constellations they lie. These are listed in the sidebar on the opposite page, along with the positions of the equinoxes and solstices.

The Moon is easy to find by its size and distinctive shape. Crescent moons are found near the Sun, while gibbous moons are found more nearly opposite the Sun. Waxing moons are found east of the Sun, and waning moons are found west of the Sun. The Moon changes its position gradually, moving eastward through the stars, about 13° per day.

Planets usually appear as bright stars. The brightness of each planet varies as our distance from it and its phase changes. The outer three planets (Uranus, Neptune, and Pluto) are difficult or impossible to see without a telescope, but the others (Mercury, Venus, Mars, Jupiter, and Saturn) can be among the most prominent objects in the sky.

Mercury and Venus always remain relatively close to the Sun in the sky and must be viewed during or close to twilight. They will be found in the zodiac, east or west of the Sun, depending on their current configurations.

FIGURE 9.3

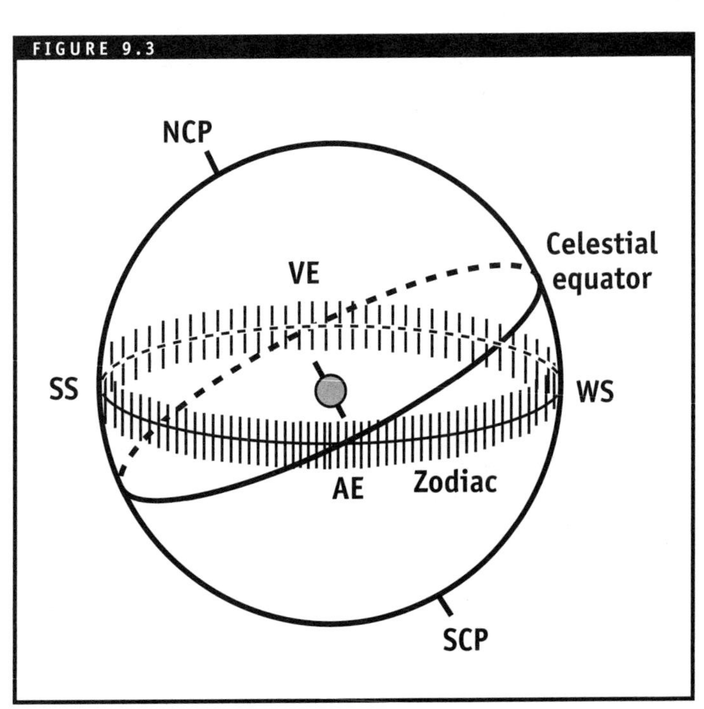

Mars, Jupiter, and Saturn travel slowly around the zodiac and may appear anywhere along it, with respect to the Sun. Their locations will also depend on their current configurations.

Planets are not visible every night. Like the Moon, they spend half the time below our horizon. In addition, when they are near conjunction, they cannot be seen due to the glare of the Sun, much as the new moon is not seen.

Locations of the planets and the Moon in its different phases can be found in each month's issue of *Sky and Telescope* or *Astronomy*. Usually the planets' positions will be marked on star charts of the equatorial regions of the celestial sphere; they will appear in the sky as bright "stars" in the zodiacal constellations, where no stars are marked on regular star charts.

Precession

Until now, we have considered the Earth's axis to have a fixed orientation in space, always pointed to the north celestial pole, near Polaris. For most casual observing, this picture is accurate enough. However, the Earth's axis is not fixed in space—it slowly changes its alignment, with the north celestial pole making a huge circle on the celestial sphere every 26,000 years, as shown in Figure 9.4.

This motion is called **precession**; it can be demonstrated by spinning a gyroscope or a top and observing the changing position of the spin axis. Just as the axis of the spinning top points to different locations on the ceiling as it precesses, the Earth's axis points to different locations on the celestial sphere. The north celestial pole, which today is about 0.9° away from the North Star, is currently getting closer to Polaris and will pass within about 0.5° of it in another hundred years or so. Following that, the NCP will drift farther away from Polaris, making it a "poorer" North Star as time goes on.

FIGURE 9.4

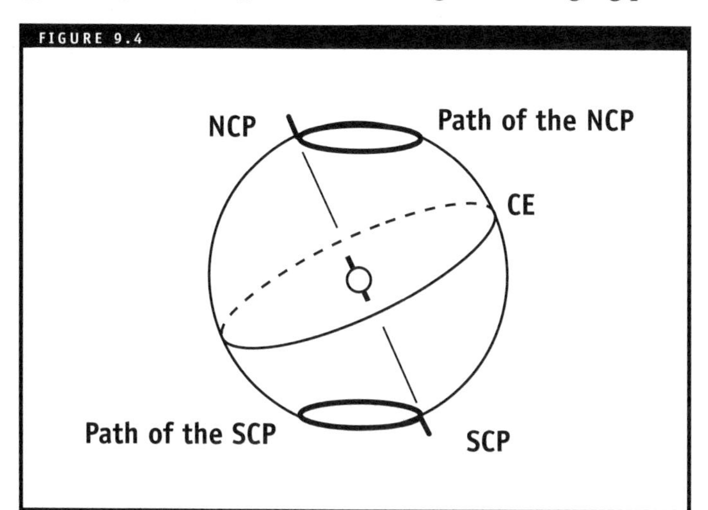

Some people think that the North Star is the brightest star in the sky; it is not—it is just the closest bright star to the NCP. Polaris has not always been our North Star. In the time of the ancient Egyptians, the NCP was quite close to Thuban (α Draconis), making it the North Star. In another 12,000 years or so, the North Star will be the very bright Vega (α Lyrae), which will be an excellent northern beacon.

As the celestial poles describe circles on the celestial sphere, the position of the celestial equator changes among the stars. This motion causes the equinoxes—the intersections of the celestial equator and the ecliptic—to drift slowly westward around the ecliptic. It is this motion of the equinoxes with respect to the stars that causes the tropical year to be slightly shorter than the sidereal year.

Precession causes no noticeable changes over a short length of time. But over several hundred years, the changes become significant, and over several thousand years they become dramatic. Recall that our seasons are determined by the tilt of the Earth's axis toward or away from the Sun.

ecliptic
the Sun's apparent path on the celestial sphere; also, the plane of the Earth's orbit; the ecliptic is not in the same plane as the celestial equator, but inclined to it by about $23^{1}/_{2}$ degrees; this angle is called the obliquity, or the tilt

precession
the slow wobble of the Earth's rotational axis and the gradual change in the positions of the equinoxes, solstices, and celestial poles that it causes

Zodiacal Constellations

- Pisces the Fishes
- Aries the Ram
- Taurus the Bull
- Gemini the Twins
- Cancer the Crab
- Leo the Lion
- Virgo the Virgin
- Libra the Scales
- Scorpius the Scorpion
- Sagittarius the Archer
- Capricornus the Goat
- Aquarius the Water Bearer

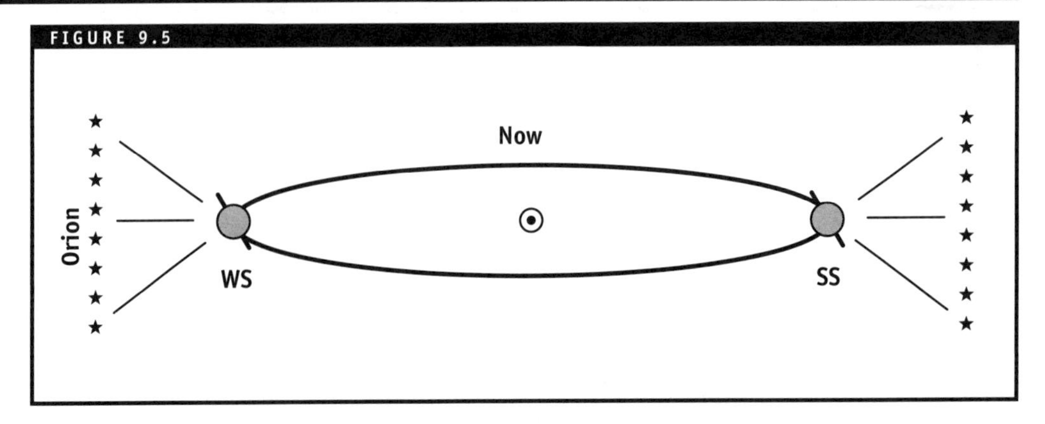

Consider the present situation as shown in Figure 9.5. In this sketch, the "winter constellations," such as Orion, would be on the left side of the celestial sphere.

Over time, precession causes the Earth's axis to shift its orientation, such that 13,000 years later it will point as shown in Figure 9.6. The stars are in approximately the same positions; Orion is still on the left side of the celestial sphere, but because the solstices have switched, Orion and the other current "winter constellations" would be viewed in the summer.

This means that any star chart will eventually become obsolete due to precession. For this reason, the better star charts identify the year, or epoch, for which they are designed (e.g., 1950 or 2000). However, for most observing needs, the popular star charts available today will serve the amateur astronomer quite well for many years to come.

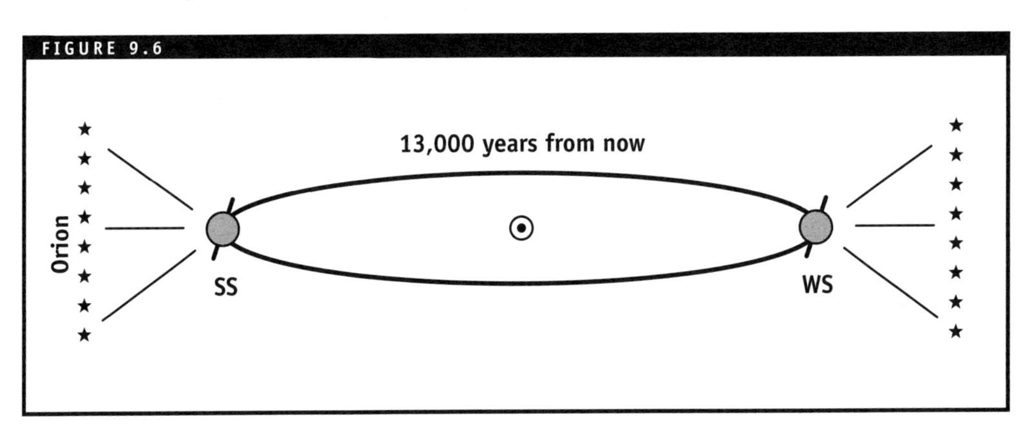

Chapter 10

Conclusion

Comet Halley 1910 PHOTO COURTESY OF NOAO

If you have arrived at the final chapter after having read all or part of this book, you will have been exposed to a fair amount of basic astronomical information. Perhaps some of it was already familiar to you; perhaps all of it was brand new; either way, you should have found some interesting tidbits which will help you understand what you see in the sky.

Most of the material in the book is unlikely to become obsolete in the near future. Future planetary missions will undoubtedly refine some of the data about the planets and their satellites, but seasons, Moon phases, eclipses, planetary configurations, and constellations are not apt to change significantly in your lifetime. Keep a pencil handy to update your copy as new moons are discovered—five were found between the writing and publishing of this edition!

The astronomy in this book is the practical, everyday, down-to-earth astronomy typically discussed in the first third of a college freshman's introductory astronomy course. That leaves plenty of areas in astronomy to discover if you are interested and have the time.

To further study astronomy, you might explore the properties of matter and learn how atoms create and interact with light. You could then understand how, by analyzing the starlight collected by their telescopes, astronomers can deduce the composition and temperature of a star. By measuring the brightness of and the distance to a star (not easily done!), they can determine the luminosity, or power output, of a star. This data permits calculation of the size of the star, while observations of binary stars—two stars orbiting each other—provide the means to find the star's mass. Most of the basic properties of stars can be determined by making the appropriate observations.

With the properties of a star and an understanding of the physical laws that govern matter and radiation, astronomers can construct a computer model of a star that yields the temperature, density, pressure, and so on at each point of its interior. Nuclear physics reveals how most stars generate enough energy by nuclear reactions to last for billions of years. Our Sun, for example, is about halfway through its ten-billion-year lifetime.

The computer models show how stars will change with time due to the gradual conversion of their hydrogen to helium by nuclear fusion. Armed with these models, astronomers can follow a star's history, from its birth at the center of a huge cloud of gas and dust pulled together by gravitational forces, through its middle age (as the Sun is now), and to its ultimate death when its nuclear fires die out. Most stars die quietly, turning into tiny white dwarfs, which gradually fade away as they cool. Others, such as the Sun, will swell up into red giants before ejecting their outer layers and forming white dwarfs from what remains. A few stars will become huge supergiants before suffering violent explosions, blowing themselves apart as supernovae and leaving behind such exotic bodies as neutron stars and black holes.

Some astronomers study the huge collection of a few hundred billion stars that we know as the Milky Way galaxy. Our Sun and essentially all of the stars that we see in the night sky are part of this galaxy, which is bound together by the mutual gravitational forces of all the matter that composes it. Our galaxy must contain many other planetary systems (we have just begun to detect some around nearby stars) and possibly other planets that harbor life—perhaps even intelligent life such as ourselves. Our galaxy is but one of billions of galaxies in the universe, which is obviously a very big place. The study of the nature, origin, and evolution of the universe is the branch of astronomy called *cosmology.*

Most of the topics mentioned in the previous paragraphs are not particularly easy to understand, although many of them are extremely interesting. All of them can be appreciated at the college freshman level, but most require some additional background in physics and mathematics for adequate comprehension. If you have the inclination, you should explore some of the rest of astronomy at an introductory level; it will certainly give you a fresh perspective on your place in the universe, which is probably the best reason to study astronomy.

To my readers who are teachers: Most students do not take an astronomy course in high school or college; the mental images and ideas of the Sun, the Moon, planets, and stars that most students carry are formed during their elementary school years. The astronomy concepts that you convey in your classroom may well be remembered for a lifetime—make the most of this opportunity.

I welcome your comments, questions, and suggestions concerning this book. Please address them to

James N. Pierce
Department of Physics and Astronomy
Minnesota State University, Mankato
141 Trafton Science Center N
Mankato, MN 56001

(email: jpierce@mnsu.edu)

Appendix

Photographing the Night Skies

Milky Way PHOTO COURTESY OF NOAO

The effect of the Earth's rotation can be seen in the motion of the sky: as the Earth turns, the sky appears to move. A camera pointed at the night sky will record star trails—streaks of light on the film, produced by stars drifting across the camera's field of view.

Star trails have a variety of instructional uses besides illustrating the motion of the night sky and its variation with direction and latitude. Very long star trails demonstrate the paths that stars follow across the sky; the curvature of the trail and the position of the star at one end or the other indicates which celestial hemisphere the star inhabits. Very short star trails created by short exposures can be used to identify constellations. Star trails of any length can serve as a backdrop for photographing meteors. Color film shows the natural colors of the stars, from which astronomers can deduce their temperatures. Best of all, star trails are a unique form of photographic art—and you can create them yourself with clear skies, a little time, and minimal equipment.

A Guide to Taking Photos of Star Trails

You will need

- a 35mm camera with a manual shutter that has a B or Bulb setting
- a locking cable release (practice using this apparatus before you take it out for night pictures)
- a tripod or other support for the camera
- color slide film (slides are better than prints, as most automatic processors will not print dark pictures; any speed can be used; ambient lighting will be held in check better by slow film, but fainter objects can be recorded with faster film; a small-diameter lens requires faster film to record the fainter stars)
- a clear night
- a dark sky, free from streetlights, car headlights, and moonlight

You may also want to have with you

- a watch for timing exposures
- a notebook and pen for recording camera settings
- a flashlight to see what you are recording
- warm clothing or mosquito repellent, depending on the season

Procedure:

1. Load your camera with film.
2. Set the camera on the tripod and point it at a group of bright stars. Take off the lens cap.
3. Set the aperture wide open to the lowest f number, usually about f/1.4 or f/1.8 for a 50mm lens. If ambient lighting is strong, stop the lens down to f/2.8 or so.
4. Set the focus to infinity (∞).
5. Set the shutter speed to B. The shutter will remain open as long as the button is pressed down.
6. Install the locking cable release on the shutter release button.
7. Cock the shutter and check the viewer for proper aiming. Don't expect to see very much through the viewer.
8. Push the cable release button and lock it down. You should hear the shutter click open. Note the starting time.
9. Wait patiently. If the skies are dark, you might wait several hours. If in town, try five minutes. Experiment with exposure times to get the desired effect. If you stroll around in the dark while you wait, be careful not to trip over the tripod. (Note: Do not bring a large dog or a small active child on this photo shoot!)
10. Unlock the cable release; you should hear the shutter click shut. (If you do not hear a click, your cable release did not lock, or your shutter was not set on B.)
11. Record the date, time, location, film speed, lens, aperture, exposure time, direction of target, and its name if you know your constellations. Then find a new target and do it all over again.
12. When you finish a roll of film, be careful in getting your film developed. Some developers will note your dark pictures with dots or streaks and assume you are just another photographic loser. They may damage or destroy your entire night's work. Before turning in a whole roll of star trails, you should try one or two shots on a roll of regular pictures to see how the developer will respond. (I still do this anyway—24 pictures of star trails require a lot of effort.)
13. When you get your pictures back, record on each slide, or on the back of each print, the date, film speed, lens, and so on. As your collection grows, you'll have the pertinent information along with each picture.

Glossary

acceleration
the rate of change of velocity of an object

analemma
the figure-8 shaped path traced over the course of a year by the Sun at a particular time of day

annular solar eclipse
a solar eclipse in which the Sun's disk appears as a ring around the new moon

annulus
ring shape

Antarctic Circle
$66\frac{1}{2}°$ south latitude; the northernmost points in the southern hemisphere where the Sun can remain above (or below) the horizon for 24 hours at certain times

aphelion
the point in an object's orbit around the Sun where it is farthest from the Sun

apogee
the point in an object's orbit around the Earth where it is farthest from the Earth

Arctic Circle
$66\frac{1}{2}°$ north latitude; the southernmost points in the northern hemisphere where the Sun can remain above (or below) the horizon for 24 hours at certain times

asteroid
small, rocky chunk of matter orbiting the Sun; also called a *minor planet;* most asteroids orbit in the asteroid belt

asteroid belt
a band between Mars and Jupiter where most asteroids orbit

astronomical unit (AU)
equal to about 93 million miles (150 million kilometers), the mean distance of the Earth from the Sun

atmosphere
layer of gases surrounding an astronomical body

autumnal (or September) equinox
the point on the ecliptic where the Sun crosses the celestial equator from north to south; marks the start of autumn in the northern hemisphere

axis
the line on which a rotating body turns

barycenter
the center of mass in a system of two orbiting bodies

celestial equator
all points on the celestial sphere that are equidistant from both celestial poles

celestial pole
one of two points that mark the intersection of Earth's rotational axis with the celestial sphere; the north celestial pole (NCP) is directly above the North Pole while the south celestial pole (SCP) is directly above the South Pole

celestial sphere
a huge, imaginary sphere around the Earth, used to model the positions and motions of celestial bodies

centrifugal force
a reaction force to a centripetal force

centripetal force
the force acting upon a body moving along a curved path that is directed toward the center of curvature of the path and constrains the body to the path

circumference
the distance around a circle or sphere

circumpolar stars
stars that never rise or set because their star trails do not intersect the horizon

comet
mass of frozen gases revolving around the Sun, generally in a highly eccentric orbit

configuration
the relative position of bodies in space

conjunction
configuration in which two objects appear close together in the sky, at or near 0° elongation

constellation
a group of stars; also, a region of the celestial sphere

crescent moon
a phase between new and quarter that shows a surface that is less than half illuminated

diameter
any line that joins two points of a circle and passes through its center or that joins two points of the surface of a sphere and passes through its center

diurnal motion
daily motions of the Earth and sky, associated with the rotation of the Earth

Earth radius
the radius of the Earth $\approx$ 6,400 km

earthshine
sunlight reflected from Earth that illuminates the dark face of a young crescent moon

eastern quadrature
configuration in which a planet has elongation = 90° E; will be highest in the sky at sunset and visible in the evening sky; will have a gibbous phase

eccentricity
the degree of flattening of an ellipse

eclipse
the darkening of a celestial body as it passes through the shadow of another body; the obscuration of all or part of the sun by a celestial body

eclipse season
a month-long time interval when eclipses occur at new or full moons; normally, two eclipse seasons occur each year, about six months apart

eclipse track
the path of the Moon's shadow along the surface of the Earth during a solar eclipse; people living along the eclipse track will see a solar eclipse when the shadow passes over them

ecliptic
the Sun's apparent path on the celestial sphere; also, the plane of the Earth's orbit; the ecliptic is not in the same plane as the celestial equator, but inclined to it by about $23\frac{1}{2}$ degrees; this angle is called the obliquity, or the tilt

ellipse
a geometric curve in the shape of a flattened circle; the shape of planetary orbits

elongation
the angle between the Sun and a planet, as seen from Earth, generally measured in degrees

equator
all points on the Earth's surface that are equidistant from both poles

equinox
the time when the Sun crosses the plane of the Earth's equator, making night and day of approximately equal length all over the world

fixed star
a star that occupies a fixed point on the celestial sphere

foci
plural of *focus* (one focus, two foci)

focus
one of two fixed points used to create an ellipse

force
a push or pull

friction
surface resistance to relative motion, as of a body sliding or rolling

galaxy
an assembly of gas, dust, and typically billions of stars, all bound together by gravity

gas
matter composed of atoms fairly independent of each other; has a free volume and a free shape

gibbous moon
a phase between quarter and full that shows a surface that is more than half illuminated

gravity
the attractive force that acts between all objects in the universe; on Earth, it is the force that pulls us down and holds us onto the planet

greatest eastern elongation
configuration in which a planet is as far east of the Sun as it can get; it rises after the Sun, follows the Sun as it moves westward across the sky and sets after the Sun sets; this planet is best viewed just after sunset in the evening sky; the planet will be only half illuminated

greatest elongation
the maximum angle from the sun

greatest western elongation
configuration in which a planet is as far west of the Sun as it can get; it rises before the Sun, precedes the Sun westward across the sky, and sets before the Sun sets; planets at this elongation are best viewed just before sunrise in the morning sky; the planet will be only half illuminated

greenhouse effect
process by which a planetary atmosphere absorbs outgoing radiation, thus maintaining higher temperatures on the surface of the planet

horizon
the circle around you where the sky and the ground meet

inferior conjunction
configuration in which a planet is in line with the Sun between the Earth and the Sun

inferior planets
planets with orbits inside the Earth's—Mercury and Venus

kilogram
a metric unit of mass approximately equal to the mass of 1000 cubic centimeters of water; on the surface of the Earth a mass of one kilogram has a weight of about 9.8 newtons and about 2.2 pounds

liquid
matter composed of particles bound weakly together; has a fixed volume and a free shape

lunar eclipse
an eclipse in which parts of the Moon are darkened by the Earth's shadow; lunar eclipses occur at times of the full moon

maria
huge lava-filled basins on planetary surfaces

mass
a measure of the amount of matter contained in a body

matter
the substance or substances of which any physical object consists, or is composed

meridian
an imaginary line that runs from the north celestial pole (NCP) through the zenith to the south celestial pole (SCP), dividing the observer's sky into eastern and western halves

meteor
the luminous trail of heated air produced by a meteoroid's passage through the Earth's atmosphere

meteorite
a meteroid that reaches the surface of the Earth or another planet or moon

meteoroid
a rock in space on a collision course with the Earth; sources of meteroids include comets and colliding asteroids

meteor shower
an event caused by the Earth's passage through the orbit of a comet, where it will collide with an increased number of meteoroids; it happens at the same point in the Earth's orbit—and thus, on the same date every year

midnight sun
the phenomenon of the Sun's remaining above the horizon at midnight (and throughout a 24-hour period); occurs only for observers north of the Arctic Circle or south of the Antarctic Circle

minor planet
an asteroid

nadir
the point in the celestial sphere directly beneath the observer, opposite the zenith

newton
a metric unit for measuring weight and other forces; the amount of force required to give a mass of one kilogram an acceleration of one meter per second per second ($1\ m/s^2$)

node
the points where the moon passes through the Earth's orbital plane

north celestial pole (NCP)
marks the intersection of Earth's rotational axis with the celestial sphere directly above the North Pole

North Pole
the northern point where the Earth's rotational axis intersects its surface

oblate
flattened at the poles

obliquity (tilt)
the angle between a planet's rotational axis and its orbital axis

opposition
configuration in which a planet is on the opposite side of the Earth from the Sun; a planet in this position behaves like a full moon, being fully illuminated and visible all night long; this is the best configuration for viewing a superior planet

partial lunar eclipse
an eclipse in which only part of the Moon enters the umbra and is darkened

partial solar eclipse
an eclipse in which some of the Sun is blocked from view by the Moon

penumbra
the light shadow outside the dark shadow of an opaque body, where the light from the source of illumination is partially cut off

penumbral lunar eclipse
an eclipse in which the Moon passes through the penumbra and appears to be shaded across its disk, not significantly darkened

perigee
the point in an object's orbit around the Earth where it is closest to the Earth

perihelion
the point in an object's orbit around the Sun where it is closest to the Sun

perpendicular
meeting at right (90°) angles

pi (π)
the ratio of the circumference of a circle to its diameter, approximately ($\approx$) 3.141592654359

planet
body in orbit around the Sun; nine planets orbit the Sun, including Earth

position
location with respect to a reference point

pound
a British unit for measuring weight and other forces; the amount of force required to give a mass of one slug an acceleration of one foot per second per second (1 ft/s^2)

precession
the slow wobble of the Earth's rotational axis and the gradual change in the positions of the equinoxes, solstices, and celestial poles that it causes

quadrature
configuration of a superior planet in which its elongation is 90°; at eastern quadrature the planet is 90° east of the Sun, while at western quadrature the planet is 90° west of the Sun

radiant
the point in the sky from which a meteor shower seems to radiate

radii
plural of *radius* (one radius, two radii)

radius
the distance from the center of a circle to a point on the circle (in a sphere, the distance from the center of the sphere to a point on the surface of the sphere)

revolution
the motion of a body that is orbiting (revolving) around another body

revolution period
the time it takes a body to complete one orbit

rings
systems of many tiny particles orbiting the planet, forming a planar ring

rotation
the motion of a body that is spinning (rotating) on an axis through the body

satellite
natural—a moon that orbits a planet
artificial—a human-made object sent into space to orbit a planet

scientific notation
number expressed as a product of two factors; the first factor is a number between 1 and 10, and the second factor is a power of 10

sidereal day
the interval between successive alignments of the Earth's axis, the observer, and a fixed star—about $23^h56^m4^s$

sidereal month
the interval between successive alignments of the Earth, the Moon, and a fixed star—about 27.32 days

sidereal time
the system of time based on the Earth's rotation in relation to fixed stars; because one sidereal day is $23^h56^m4^s$ long, sidereal time flows more rapidly than solar time

sidereal year
the interval between successive alignments of the Earth, the Sun, and a fixed star—about 27.32 days

slug
a British unit of mass equivalent to approximately 14.6 kilograms; on the surface of the Earth a mass of one slug has a weight of about 32.2 pounds

solar corona
a faint halo of gases surrounding the Sun visible during a total solar eclipse

solar day
the interval between successive alignments of the Earth's axis, the observer, and the Sun—24 hours

solar eclipse
an eclipse in which parts of the Sun are hidden from view by the Moon; solar eclipses occur at times of the new moon

Solar System
the Sun and all of the objects that revolve around it

solid
matter composed of tightly bound atoms; has a fixed volume and a fixed shape

south celestial pole (SCP)
marks the intersection of Earth's rotational axis with the celestial sphere directly above the South Pole

South Pole
the southern point where the Earth's rotational axis intersects its surface

speed
the distance covered per unit of time by an object in motion

star
one of many points of light in the night sky that maintains a fairly constant position with respect to its neighbors; a massive, gaseous sphere, heated by gravitational compression until it radiates visible light; most stars generate energy by thermonuclear fusion; the Sun is a very close star

star chart
a map of the celestial sphere showing the locations of fixed stars

star trails
streaks made by stars on photographs of the sky, caused by Earth's rotation

summer (or June) solstice
the northernmost point on the ecliptic where the Sun is directly above the tropic of Cancer; marks the beginning of summer in the northern hemisphere

superior conjunction
configuration in which a planet is in line with the Sun on the opposite side of the Sun from the Earth

superior planets
planets with orbits lying outside the Earth's—Mars, Jupiter, Saturn, Uranus, Neptune, and Pluto

surface area
the total area of the surface of a solid

synodic month
measures the time required for the Moon to orbit once and line up again with the Earth and the Sun—about 29.53 days

synodic period
the time it takes for an inferior planet to lap the Earth or for the Earth to lap a superior planet

temperature
a measure of the amount of energy contained in a body

thermal lag
the time required for an object to heat up or cool down in response to changes in the rate of incoming heat

total lunar eclipse
an eclipse in which the Moon is completely immersed in the Earth's umbra, where practically no light can reach it

total solar eclipse
an eclipse in which all of the Sun's disk is blocked from view by the Moon

tropical year
the interval between successive alignments of the Earth, the Sun, and the vernal equinox—$365^{d}5^{h}48^{m}46^{s}$

tropic of Cancer
the latitude that marks the northern limit at which the Sun can be seen at the zenith, about $23\frac{1}{2}$°N; the Sun is overhead at the tropic of Cancer on the summer solstice

tropic of Capricorn
the latitude that marks the southern limit at which the Sun can be seen at the zenith, about $23\frac{1}{2}$°S; the Sun is overhead at the tropic of Capricorn on the winter solstice

umbra
the dark shadow of an opaque body, where direct light from the source of illumination is completely cut off

universe
the totality of known or supposed objects and phenomena throughout space

vector
a physical quantity that is described by both magnitude and direction

velocity
speed in a specific direction

vernal (or March) equinox
the point on the ecliptic where the Sun crosses the celestial equator from south to north; marks the start of spring in the northern hemisphere

volume
the amount of space that an object occupies

waning moon
any phase during which the illuminated fraction of the Moon's face decreases (from full to last quarter to new)

waxing moon
any phase during with the illuminated fraction of the face increases (from new to first quarter to full)

weight
a force; the magnitude of which depends upon the mass of the object and the gravity exerted upon it

western quadrature
configuration in which a planet has elongation = 90°W; will be highest in the sky at sunrise and visible in the morning sky; will have a gibbous phase

winter (or December) solstice
the southernmost point on the ecliptic where the Sun is directly above the tropic of Capricorn; marks the start of winter in the northern hemisphere

zenith
the point in the sky directly above the observer, opposite the nadir

Index